W0257448

ERGEBNISSE DER MATHEMATIK UND IHRER GRENZGEBIETE

UNTER MITWIRKUNG DER SCHRIFTLEITUNG DES
„ZENTRALBLATT FÜR MATHEMATIK"

HERAUSGEGEBEN VON

P. R. HALMOS · R. REMMERT · B. SZÖKEFALVI-NAGY

UNTER MITWIRKUNG VON

L. V. AHLFORS · R. BAER · F. L BAUER · R. COURANT · A. DOLD
J. L. DOOB · S. EILENBERG · M. KNESER · T. NAKAYAMA
H. RADEMACHER · B. SEGRE · E. SPERNER

REDAKTION P. R. HALMOS

═══════ NEUE FOLGE · BAND 33 ═══════

REIHE:

MODERNE TOPOLOGIE

BESORGT

VON

A. DOLD und S. EILENBERG

Springer-Verlag Berlin Heidelberg GmbH
1964

DIFFERENTIABLE PERIODIC MAPS

BY

P. E. CONNER AND E. E. FLOYD

PROFESSORS OF MATHEMATICS
UNIVERSITY OF VIRGINIA
CHARLOTTESVILLE

Springer-Verlag Berlin Heidelberg GmbH
1964

© Springer-Verlag Berlin Heidelberg 1964

Originally published by Springer-Verlag OHG. Berlin . Göttingen . Heidelberg in 1964

Softcover reprint of the hardcover 1st edition 1964

Library of Congress Catalog Card Number 63-23135

ISBN 978-3-662-34309-8   ISBN 978-3-662-34580-1 (eBook)
DOI 10.1007/978-3-662-34580-1

Titel Nr. 4577

Preface

This research tract contains an exposition of our research on bordism and differentiable periodic maps done in the period 1960—62. The research grew out of the conviction, not ours alone, that the subject of transformation groups is in need of a large infusion of the modern methods of algebraic topology. This conviction we owe at least in part to ARMAND BOREL; in particular BOREL has maintained the desirability of methods in transformation groups that use differentiability in a key fashion [9, Introduction], and that is what we try to supply here.

We do not try to relate our work to Smith theory, the homological study of periodic maps due to such a large extent to P. A. SMITH; for a modern development of that subject which expands it greatly see the BOREL Seminar notes [9]. It appears to us that our work is independent of Smith theory, but in part inspired by it. We owe a particular debt to G. D. MOSTOW, who pointed out to us some time ago that it followed from Smith theory that an involution on a compact manifold, or a map of prime period p on a compact orientable manifold, could not have precisely one fixed point. It was this fact that led us to believe it worthwhile to apply cobordism to periodic maps.

On the personal side we are greatly endebted to DEANE MONTGOMERY and G. T. WHYBURN, who have supported our work in every way possible. We would also like to thank JIN-CHEN SU, who has read critically some of the manuscript. During portions of the time spent on the research presented here, we have been supported by AFOSR Grant AF 49(638)-72 and NSF Grant G18853. We have also received generous support from the Sloan Foundation as Sloan Fellows.

P. E. CONNER and
E. E. FLOYD
Charlottesville, Virginia

November 5, 1962

Introduction

In this introduction we speak of a compact differentiable n-manifold as simply an n-manifold. The boundary of V^n is denoted by $\dot{V}^n$; V^n is closed if $\dot{V}^n = \emptyset$. Two closed n-manifolds V^n and V'^n are in the same bordism class if there exists an $(n + 1)$-manifold W^{n+1} with $\dot{W}^{n+1}$ the disjoint union of V^n and V'^n (we adopt here ATIYAH's suggestions on the usage of "bordism" and "cobordism"). There results the abelian group $\mathfrak{N}_n$ of (unoriented) bordism classes, due to THOM [40] and completely computed by him. Moreover $\mathfrak{N} = \sum \mathfrak{N}_n$ is a ring with multiplication induced by the cartesian product. THOM has shown that the structure of $\mathfrak{N}$ is that of a polynomial algebra, over the base field Z_2, with a generator in each dimension not of the form $2^j - 1$.

THOM also considered the bordism classes of pairs (V^n, o) where V^n is a closed n-manifold and o is an orientation of V^n. There resulted the oriented bordism groups Ω_n, and the bordism ring $\Omega = \Sigma\, \Omega_n$. THOM computed the rank of the free part of Ω_n; the torsion has since been settled by the work of MILNOR [26] and WALL [42]. In particular, Ω has no torsion of odd order (MILNOR) and the 2-torsion consists entirely of elements of order two (WALL). Moreover Ω/Tor, where Tor denotes the torsion subgroup, is polynomial algebra with a generator in each dimension $4k$ (MILNOR).

Various refinements of Ω (and of $\mathfrak{N}$) result by putting additional structure on the manifold V^n; this has occurred to several people, in particular ATIYAH [1], EELS, and MILNOR. The main point of this tract is that such refinements are particularly appropriate in the study of periodic maps and transformation groups. The role of Chapters I and II is to present the purely topological aspects of our work.

Let $X \supset A$ denote a pair of spaces. An oriented singular n-manifold in (X, A) is a pair (V^n, f) where V^n is an oriented n-manifold and $f: (V^n, \dot{V}^n) \to (X, A)$ is a map. There is a suitable bordism relation joining such pairs (V^n, f), and there results an abelian group $\Omega_n(X, A)$; denote the bordism class of (V^n, f) by $[V^n, f]$. Given (V^n, f) and a closed oriented m-manifold M^m, there is the pair $(V^n \times M^m, f\pi)$ where $\pi: V^n \times M^m \to V^n$ is projection. We consider $\Omega_*(X, A) = \Sigma\, \Omega_n(X, A)$ as an Ω-module by

$$[V^n, f]\,[M^m] = [V^n \times M^m, f\pi]\,.$$

Maps $\varphi: (X, A) \to (Y, B)$ induce homomorphisms $\varphi_*: \Omega_n(X, A) \to \Omega_n(Y, B)$; there is also a boundary homomorphism $\partial: \Omega_n(X, A) \to \Omega_{n-1}(A)$.

In Chapter I, we study $\{\Omega_*(X, A),\ \varphi_*,\ \partial\}$ as a generalized homology theory; such a study has also been carried out by ATIYAH [1]. All the Eilenberg-Steenrod axioms for a homology theory are satisfied except for the "dimensional axiom"; for a point p, $\Omega_n(p)$ is the Thom group Ω_n. We also present a bordism spectral sequence, a spectral sequence whose E^∞-terms are associated with a filtration of $\Omega_*(X, A)$ and which has $E^2_{p,q} = H_p(X, A;\ \Omega_q)$.

We also give in Chapter I a homotopy interpretation for $\Omega_n(X, A)$. Here the Thom spaces $MSO(k)$ enter, with $MSO(k) = E/\dot{E}$ where E is the bundle space of a universal k-cell bundle $E \to BSO(k)$. Recall that THOM proved that $\Omega_n \cong \pi_{n+k}(MSO(k))$ for $k \geq n + 2$. We extend THOM's result to show that

$$\Omega_n(X, A) \cong \pi_{n+k}((X/A) \wedge MSO(k)), \quad k \geq n + 2,$$

where $\wedge$ denotes the smashed product. Thus the bordism functor $\Omega_*(X, A)$ is shown to fit into G. W. WHITEHEAD's generalized homology theory [43].

In Chapter II we show that $\Omega_*(X, A)$ can be computed for a wide range of CW pairs (X, A). The chief tool of the section is that the bordism spectral sequence is trivial modulo the class C of odd torsion groups; the proof relies heavily on WALL's result on the structure of $H^*(MSO; Z_2)$ as a module over the Steenrod algebra. Moreover

$$\Omega_n(X, A) \cong \Sigma_{p+q=n} H_p(X, A;\ \Omega_q) \bmod C.$$

If $H_*(X, A; Z)$ has no odd torsion, then

$$\Omega_n(X, A) \cong \Sigma_{p+q=n} H_p(X, A;\ \Omega_q).$$

For finite CW pairs (X, A) with $H_*(X, A; Z)$ torsion-free, then

$$\Omega_*(X, A) \cong H_*(X, A; Z) \otimes \Omega.$$

Given a closed oriented manifold V^n, there are the Whitney numbers of V^n and the Pontryagin numbers of V^n; these are functions of the bordism class $[V^n]$. We generalize these in Chapter II to obtain Whitney numbers and Pontryagin numbers of a map $f: V^n \to X$. These turn out to depend only on the bordism class of $f: V^n \to X$. Moreover if the torsion of $H^*(X; Z)$ consists of elements of order two, these numbers determine the bordism class $[V^n, f]$.

Along with all the above there is also an unoriented theory $\{\mathfrak{N}_*(X, A),\ \varphi_*,\ \partial\}$, in which V^n is not required to be oriented. As with ordinary bordism, this is an easier case. For example, for all CW pairs it is the case that

$$\mathfrak{N}_*(X, A) = H_*(X, A; Z_2) \otimes \mathfrak{N}.$$

In Chapter III we begin to make the transition to transformation groups. For G a finite group, consider all pairs (G, V^n) consisting of a closed oriented n-manifold V^n and a free action of G on V^n which preserves the orientation. Two such (G, V^n) and (G, V'^n), are in the same bordism class if there exists an oriented $(n + 1)$-manifold W^{n+1} with $\dot{W}^{n+1} = V^n \cup - V'^n$ and a free orientation preserving action of G on W^{n+1} which yields the original action on V^n and on V'^n. Denote the resulting bordism group by $\Omega_n(G)$. It is shown that $\Omega_n(G)$ is isomorphic to the group $\Omega_n(B(G))$ of Chapter I, where $B(G)$ is a classifying space for G. There is a similar unoriented group $\mathfrak{N}_n(G) \cong \mathfrak{N}_n(B(G))$.

It is not surprising that the study of differentiable periodic maps of prime period p breaks naturally into the cases $p = 2$ and p odd. It is typical of the case $p = 2$ that orientation can be ignored, making the resulting theory simpler. In Chapters IV—VI we study the case $p = 2$; that is, we study the differentiable involutions.

First of all, we consider pairs (T, V^n) where V^n is a closed n-manifold and T is a differentiable fixed point free involution on V^n. There is a suitable bordism relation which leads to the group $\mathfrak{N}_n(Z_2)$ already discussed in Chapter III. Hence the bordism theory of fixed point free involutions is easily carried out.

We go on in Chapter IV to study the fixed point sets of differentiable involutions $T: V^n \to V^n$. Our method is based on two procedures. First, we consider a tubular neighborhood N of the fixed point set, invariant under T and of small radius. Delete from V^n the interior N^0 of N. There results an n-manifold $W^n = V^n \backslash N^0$ and a fixed point free involution $T: W^n \to W^n$. Hence the pair $(T|\dot{N}, \dot{N})$ bords in the sense of the previous paragraph, where $\dot{N}$ is the boundary both of N and of W^n. Now $\dot{N}$ is just the normal sphere bundle to the fixed point set F of T; the implications from $[T|\dot{N}, \dot{N}]_2 = 0$ can now be obtained from the fixed point free case. The other procedure gives the bordism class of V^n in terms of F and its normal sphere bundle. Take the Whitney join of the normal sphere bundle $\dot{N} \to F$ with a trivial 0-sphere bundle; there results a sphere bundle $V'^n \to F$, and V'^n has a natural involution T', the antipodal map on each fiber. It is shown that the manifolds V^n and V'^n/T' are bordant (mod 2). Appropriate combination of these two basic procedures produces a number of novel facts about fixed points of differentiable involutions. For example, given a positive integer k there exists a positive integer $\varphi(k)$ such that if $T: V^n \to V^n$ is a differentiable involution on a closed non-bording n-manifold with $n > \varphi(k)$, then some component of the fixed point set F is of dimension $> k$.

In Chapter V we discuss differentiable involutions $T_i: V^n \to V^n$, $i = 1, \ldots, k$, with $T_i T_j = T_j T_i$; that is, we discuss differentiable actions of $(Z_2)^k = Z_2 \times \cdots \times Z_2$. A stationary point of such an action is a point $x \in V^n$ with $T_i(x) = x$, all i. It is proved that if $(Z_2)^k$ acts differentiably on V^n without stationary points, then $[V^n]_2 = 0$. It follows that in any differentiable action of $(Z_2)^k$ on a closed manifold V^n, then the set F of stationary points, together with the normal bundle to F and the action of $(Z_2)^k$ on the normal bundle, determine the bordism class $[V^n]_2$. Since this appears very difficult to cope with in general, we content ourselves with a single special case. We consider actions of $Z_2 \times Z_2$ on closed manifolds V^n with all stationary points isolated. Among other things it then follows that $[V^n]_2 = 0$ or $[V^n]_2 = [P_2 \times \cdots \times P_2]$ where P_2 is the real projective plane.

Chapter VI is concerned with an operation on bundles. Namely given a fixed point free involution (T, B) and a bundle $r : E \to B/T$ of n-dimensional vector spaces, there is defined another n-dimensional vector space bundle $r : E \to B/T$, which we call the *twist* of r by (T, B). It is shown to be a particularly simple case of the tensor product. One application is concerned with an m-dimensional component F^m of the fixed point set of a differentiable involution $T : V^{m+n} \to V^{m+n}$. Namely if $H^i(V^{m+n}; Z_2) = 0$ for $n - k \leq i \leq n$ then the Whitney classes V_i of the normal bundle to F^m are trivial for $n - k \leq i \leq n$.

A second application of the twist construction of Chapter VI consists of generalizing the Borsuk antipode theorems. Suppose that f is a map of S^n into the differentiable manifold M^n. If f is of even degree (that is, if $f^* : H^n(M^n; Z_2) \to H^n(S^n; Z_2)$ is trivial), then there exists $x \in S^n$ with $f(-x) = f(x)$. There is also a discussion of maps $f : S^n \to M^k$, $k < n$.

In Chapter VII we consider the structure of $\Omega_*(Z_p)$, p an odd prime. Alternatively we consider the group of bordism classes of pairs (T, V^n) where V^n is a closed oriented manifold and T is an orientation preserving, fixed point free diffeomorphism on V^n of period p. The bordism spectral sequence of $B(Z_p)$ is trivial; thus $\Omega_n(Z_p)$ is determined up to group extensions. The main task of Chapter VII is to determine the precise additive structure of $\Omega_*(Z_p)$. This is carried out; the generators of $\tilde{\Omega}_*(Z_p)$ as an Ω-module turn out to be pairs (T, S^{2n-1}), one for each positive integer n, and these have order p^{a+1} where $a(2p - 2) < 2n - 1 < < (a + 1)(2p - 2)$. We also discuss $\Omega_*(Z_p k)$.

Chapter VIII turns to a discussion of fixed point sets F of differentiable maps $T : V^n \to V^n$ of odd prime period; here it is assumed that V^n is oriented and that T preserves orientation. The first task is to understand the normal bundle to F. It is shown that the structural group of the normal bundle can always be reduced to the unitary group. We then turn to the following problem, whose solution appears to make use of all our technique up to this point; which bordism classes of Ω_n admit representatives V^n upon which $Z_p \times Z_p$ acts, preserving the orientation and without stationary points? It is clear that $Z_p \times Z_p$ so acts on $Y^0 = p$

points; BOREL has pointed out that it also acts on complex projective space $Y^{2p-2} = P_{p-1}(C)$ with

$$T_1[z_1, \ldots, z_p] = [z_2, \ldots, z_p, z_1], \qquad T_2[z_1, \ldots, z_p] = [z_1, \varrho z_2, \ldots, \varrho^{p-1} z_p]$$

where $\varrho = \exp(2\pi i/p)$. We prove that if $Z_p \times Z_p$ acts differentiably on V^n, preserving the orientation and without stationary points, then $[V^n]$ is in the ideal of Ω generated by $[Y^0]$ and $[Y^{2p-2}]$.

We also discuss in Chapter VIII the structure of the manifold V^n and the fixed point set F, in the case where $T: V^n \to V^n$ has the normal bundle to the fixed point set trivial in a suitably strong sense.

In Chapter IX we turn to differentiable actions of $(Z_p)^k$. Here our work is motivated by that of BOREL [6, 9]. In view of his work together with our point of view of the previous chapters, it is natural to ask the following question: which bordism classes of Ω admit representatives upon which $(Z_p)^k$ acts preserving the orientation and without stationary points? The collection of all such bordism classes constitutes an ideal $SF((Z_p)^k)$ in Ω. We are not able to compute this ideal in general, but we make some progress in determining its structure. For example, it is contained in the ideal of all $[V^n]$ with the Pontryagin numbers of $[V^n]$ all divisible by p. While we are about it, we consider also actions of abelian groups of order p^k.

CHAPTER I

The bordism groups

Given a pair (X, A) of spaces, consider all maps $f: (B^n, \dot{B}^n) \to (X, A)$, where B^n is a compact oriented differentiable manifold with boundary $\dot{B}^n$. We introduce a relation of the bordism type on the class of all such f, arriving in § 4 at the set $\Omega_n(X, A)$ of equivalence classes. These groups constitute a generalized homology theory; it is shown in § 5 that the Eilenberg-Steenrod axioms are satisfied except for the dimensional axiom. In § 12 we give a homotopy interpretation for $\Omega_n(X, A)$. It is shown that $\Omega_n(X, x_0)$, X a CW complex with base point $x_0 \in X$, is given by the homotopy group $\pi_{n+k}(X \wedge MSO(k))$; here $MSO(k)$ is the Thom space and $k \geq n + 2$. In § 13 the dual generalized cohomology theory is sketched; this is due to ATIYAH [1]. We try to fill in along the way some of the background material of differential topology.

1. Differentiable manifolds

In this section we outline the elementary properties of differentiable manifolds, paying particular attention to manifolds with boundary. Always in this work "differentiable" is used as an abbreviation for "differentiable of class C^∞".

A map $f: A \to R^n$, where A is a subset of Euclidean m-space R^m, is said to be *differentiable* if and only if for each $x \in A$ there is a neighborhood V of x in R^m and a differentiable map $F: V \to R^n$ with $F(x) = f(x)$ for $x \in V \cap A$.

Denote by H^n the closed half-space in R^n consisting of all points $(x_1, x_2, \ldots, x_n)$ with $x_1 \geq 0$. A separable metric space B^n is a *topological n-manifold* if for each $x \in B^n$ there is a neighborhood V of x and a homeomorphism h of V onto an open subset of H^n. A *differentiable structure* on a topological n-manifold B^n consists of a collection of ordered pairs (V_i, h_i), where V_i is open in B^n and h_i is a homeomorphism of V_i onto an open subset of H^n, such that

   i) the collection $\{V_i\}$ is an open covering of B^n;

   ii) for every pair (i, j), the map $h_j h_i^{-1}: h_i(V_i \cap V_j) \to R^n$ is differentiable;

   iii) the collection $\{(V_i, h_i)\}$ is maximal with respect to the above properties.

Given a collection of pairs satisfying i) and ii), there is a unique collection of pairs containing it and satisfying i)—iii). A topological n-manifold B^n together with a differentiable structure on B^n is a *differentiable n-manifold*. The pairs (V_i, h_i) are referred to as *coordinate neighborhoods*.

In the above definitions, H^0 is taken to be a single point; that is a 0-manifold is a countable discrete space.

Denote by $R^{n-1} \subset H^n$, $n > 0$, the set of all $(x_1, \ldots, x_n)$ with $x_1 = 0$. The boundary $\dot{B}^n$ of a differentiable manifold consists of those points $x \in B^n$ for which there is a coordinate neighborhood (V, h) with $x \in V$ and $h(x) \in R^{n-1} \subset H^n$. If $\dot{B}^n$ is empty, then B^n is a *manifold without boundary*; in this case H^n may be replaced by R^n in the definition of differentiable structure. Always the boundary $\dot{B}^n$ of a differentiable n-manifold is a differentiable $(n-1)$-manifold without boundary, such that whenever (V, h) is a coordinate neighborhood of B^n then $(V \cap \dot{B}^n, h \,|\, V \cap \dot{B}^n)$ is a coordinate neighborhood of $\dot{B}^n$.

If U is an open subset of a differentiable n-manifold B^n, then U is also a differentiable n-manifold. Namely, its coordinate neighborhoods are the coordinate neighborhoods (V, h) of B^n with $V \subset U$. We call this the differentiable structure induced by that of B^n. The following easily verified remark is useful in piecing together structures.

(1.1) *Suppose that U_1 and U_2 are open subsets of the topological n-manifold B^n which cover B^n. Suppose U_1 and U_2 have differentiable structures which induce the same differentiable structure on $U_1 \cap U_2$. There exists a unique differentiable structure on B^n which induces the differentiable structure of U_1 and of U_2.*

If M^m is a differentiable manifold without boundary and B^n a differentiable manifold, the reader will see that there is an induced differentiable structure on $M^m \times B^n$. More generally, $B_1^m \times B_2^n \setminus \dot{B}_1^m \times \dot{B}_2^n$ has a natural differentiable structure.

A map $\varphi : B_1^m \to B_2^n$ connecting differentiable manifolds is *differentiable* if whenever (V, h) and (W, k) are coordinate neighborhoods of B_1^m and B_2^n respectively, then $k \varphi h^{-1} : h(V \cap \varphi^{-1}W) \to R^n$ is differentiable. The differentiable manifolds B_1^n and B_2^n are *diffeomorphic* if there exists a homeomorphism f of B_1^n onto B_2^n with both f and f^{-1} differentiable.

We assume the following differentiable collaring theorem (see MIL-NOR [27]).

(1.2) Theorem. *For any differentiable manifold B^n there is an open set U containing $\dot{B}^n$ and a diffeomorphism φ of U onto $\dot{B}^n \times [0,1)$ with $\varphi(x) = (x, 0)$ for $x \in \dot{B}^n$.*

In this work we will be mainly concerned with compact differentiable manifolds, and often with the *closed manifolds*, namely the compact differentiable manifolds without boundary. That being the case, we discuss orientability only in the compact case. Suppose that B_c^n is a component of the compact differentiable manifold B^n; the singular homology group $H_n(B_c^n, \dot{B}_c^n; Z)$ is either Z or 0 [19, p. 314]. Say that B^n is *orientable* if $H_n(B_c^n, \dot{B}_c^n; Z) \cong Z$ for each component B_c^n; to *orient* such a manifold means to select a generator σ_c of each $H_n(B_c^n, \dot{B}_c^n; Z)$. Identifying $H_n(B^n, \dot{B}^n)$ with $\Sigma_c H_n(B_c^n, \dot{B}_c^n)$ the orientation class $\sigma(B^n)$ of an oriented manifold is the element of $H_n(B^n, \dot{B}^n; Z)$ given by $\Sigma \sigma_c$.

An orientation of B^n induces an orientation of $\dot{B}^n$ by assigning the orientation $\partial \sigma(B^n)$, where $\partial : H_n(B^n, \dot{B}^n) \to H_{n-1}(\dot{B}^n)$ is the boundary homomorphism. A diffeomorphism $\varphi : B_1^n \cong B_2^n$ of oriented manifolds is *orientation preserving* if and only if $\varphi_* \sigma(B_1^n) = \sigma(B_2^n)$.

2. The Thom bordism groups

In this section we define and give elementary properties of the Thom groups Ω_n and $\mathfrak{N}_n$. At the end of the section we summarize briefly the deeper structure of these groups; however it is some time before that information is actually used. *Hereafter in this book we use "manifold" for "differentiable manifold".*

Given a closed oriented manifold M^n, then $-M^n$ is the oriented manifold obtained by using as orientation class $-\sigma(M^n)$. Given closed oriented manifolds M_1^n and M_2^n, we call the *disjoint union* of M_1^n and M_2^n any compact oriented manifold which is a disjoint union $M_1'^n \cup M_2'^n$ of closed submanifolds with $M_i'^n$ diffeomorphic to M_i^n via an orientation preserving diffeomorphism; denote a disjoint union simply by $M_1^n \cup M_2^n$.

A closed oriented manifold M^n is said to *bord* if there is a compact oriented manifold B^{n+1} with $\dot{B}^{n+1}$ diffeomorphic to M^n via an orientation preserving diffeomorphism. Two closed oriented manifolds M_1^n and M_2^n

are *bordant* if the disjoint union $M_1^n \cup -M_2^n$ bords; we refer to this relation as the *bordism* relation.

(2.1) *The bordism relation is an equivalence relation on the class of closed oriented n-manifolds. The resulting set Ω_n of equivalence classes is an abelian group with addition induced by disjoint union.*

Proof. In order to see that M^n is bordant to itself, form the oriented manifold $I \times M^n$ where I is the oriented unit interval. Then $(I \times M^n)^{\cdot} = \dot{I} \times M^n = 1 \times M^n \cup 0 \times -M^n$. Hence $M^n \cup -M^n$ bords. The bordism relation is clearly symmetric. It is in the proof of transitivity that the collaring theorem (1.2) plays its typical role. Suppose that M_1^n is bordant to M_2^n, and that M_2^n is bordant to M_3^n. There are the compact oriented manifolds B_1^{n+1}, B_2^{n+1} with $\dot{B}_1^{n+1} = M_1^n \cup -M_2^n$, $\dot{B}_2^{n+1} = M_2^n \cup -M_3^n$. We may as well suppose $B_1^{n+1} \cap B_2^{n+1} = M_2^n$. Let $B^{n+1} = B_1^{n+1} \cup B_2^{n+1}$; then B^{n+1} is a topological oriented $(n + 1)$-manifold with $\dot{B}^{n+1} = M_1^n \cup -M_3^n$. It is necessary to give B^{n+1} a suitable differentiable structure. By (1.2) there is an open set U_1 containing M_2^n in B^{n+1} and a homeomorphism h of M_1 onto $M_2^n \times (-1,1)$ with $h: U_1 \cap B_1^{n+1} \to M_2^n \times (-1,0]$ a diffeomorphism and $h: U_1 \cap B_1^{n+1} \to M_2^n \times [0,1)$ a diffeomorphism. Give U_1 the differentiable structure obtained from that of $M_2^n \times (-1,1)$ via h. Let $U_2 = B^{n+1} \smallsetminus M_2^n$; U_2 has a natural differentiable structure. We thus get a differentiable structure on B^{n+1} via (1.1). The relation is then transitive.

The equivalence class to which M^n belongs is denoted by $[M^n]$, and the collection of all such classes by Ω_n. An abelian group structure is put on Ω_n by disjoint union: $[M_1^n] + [M_2^n] = [M_1^n \cup M_2^n]$. The zero element of the group consists of those manifolds which bord; moreover, $-[M^n] = [-M^n]$.

The weak direct sum $\Omega_* = \Sigma \Omega_n$ can be given the structure of a graded ring. The product of homogeneous elements $[M_1^m]$ and $[M_2^n]$ is given by $[M_1^m] \cdot [M_2^n] = [M_1^m \times M_2^n]$. This is seen to give a well-defined associative and distributive operation. Furthermore there is a unit element, the bordism class of a single point. It is also the case that $[M_1^m]\, [M_2^n] = (-1)^{mn}\, [M_2^n]\, [M_1^m]$.

The groups Ω_n have been completely determined. THOM showed, in his original work on the subject [40], that $\Omega_* \otimes Q$, Q the rationals, is a polynomial algebra with generators the bordism classes of the complex projective spaces $P_{2k}(C)$, $k = 1, 2, \ldots$. MILNOR [25, 26], and independently AVERBUCH [2], showed that Ω_* has no elements of odd order. The torsion was then completely determined by WALL [42], who showed that all torsion consists of elements of order two, and moreover settled the structure of the set of elements of order two. Here he found useful the elements of order two discovered earlier by DOLD [16]. In the meantime, MILNOR [25, 41] had also completed THOM's work on $\Omega_* \otimes Q$ by giving the multiplicative structure of Ω_*/Tor, Tor the torsion of Ω_*. Namely there exist closed oriented manifolds Y^{4k}, $k = 1, 2, \ldots$, such that Ω_*/Tor is the polynomial algebra over the integers generated by the $[Y^{4k}]$. For $2k + 1$ a prime p, one may choose $Y^{4k} = P_{p-1}(C)$.

In addition to the oriented bordism there is also an unoriented bordism theory. In the unoriented theory all closed manifolds are used; no requirements of orientability or orientedness are imposed. The definitions of the bordism relation are otherwise precisely as in the oriented case. The unoriented bordism class of M^n is denoted by $[M^n]_2$, and the set of all bordism classes by $\mathfrak{N}_n$. As with Ω_n, $\mathfrak{N}_n$ has an abelian group structure defined by disjoint union; however every element of $\mathfrak{N}_n$ has order two. The weak direct sum $\mathfrak{N}_* = \Sigma \mathfrak{N}_n$ has a multiplication induced by the cartesian product, and $\mathfrak{N}_*$ is a graded commutative algebra over Z_2. THOM [40] completely settled the structure of $\mathfrak{N}_*$ by showing that for every n not of the form $2^t - 1$ there exists a closed manifold M^n with $\mathfrak{N}_*$ the polynomial algebra

$$Z_2([M^2]_2, [M^4]_2, [M^5]_2, \ldots) .$$

In even dimensions the M^{2k} can be chosen to be the real projective spaces $P_{2k} = P_{2k}(R)$; explicit generators were given in the odd dimensions by DOLD [16].

3. Straightening the angle

We show here that the cartesian product of two differentiable manifolds has a differentiable structure. The technique involved, usually referred to as "straightening the angle", is useful whenever two manifolds are sewn together. We follow MILNOR's exposition [21, p. 34].

Let $R_+ \subset R$ consist of all non-negative real numbers. Pick once and for all a homeomorphism τ of the quadrant $R_+ \times R_+$ onto the halfspace $R \times R_+$, with τ a diffeomorphism of $R_+ \times R_+ \backslash (0,0)$ onto $R \times R_+ \backslash (0,0)$. For example in polar coordinates let $\tau(\varrho, \theta) = (\varrho, 2\theta)$, $0 \leq \theta \leq \pi/2$.

Suppose now that we have a topological manifold B^n, and that we also have in B^n a submanifold M^{n-2} without boundary, closed in B^n, such that

   i) $B^n \backslash M^{n-2}$ has a differentiable structure;

   ii) there exist a neighborhood U of M^{n-2} in B^n and a homeomorphism Φ of U onto $M^{n-2} \times R_+ \times R_+$ with $\Phi(x) = (x, 0, 0)$ for $x \in M^{n-2}$ and with Φ a diffeomorphism of $U \backslash M^{n-2}$ onto $M^{n-2} \times R_+ \times R_+ \backslash M^{n-2} \times \times 0 \times 0$.

Let $\tau' : M^{n-2} \times R_+ \times R_+ \to M^{n-2} \times R \times R_+$ be given by $\tau'(x, y, z) = (x, \tau(y, z))$. We then have the homeomorphism $\tau' \Phi : U \to M^{n-2} \times \times R \times R_+$. There is the product differentiable structure for $M^{n-2} \times \times R \times R_+$, and hence a differentiable structure on U such that $\tau' \Phi$ is a diffeomorphism. Then U and $B^n \backslash M^{n-2}$ have differentiable structures, and these are seen to induce the same differentiable structure on their intersection. Use of (1.1) then gives a differentiable structure for B^n. This structure is referred to as obtained by *straightening the angle*.

Consider for example differentiable manifolds B_1^m and B_2^n. As pointed out in section 1, $B_1^m \times B_2^n \setminus \dot{B}_1^m \times \dot{B}_2^n$ has a natural differentiable structure. Also $\dot{B}_1^m$ and $\dot{B}_2^n$ have neighborhoods U_1 and U_2 in B_1^m and B_2^n respectively, with diffeomorphisms Φ_1 of U_1 onto $\dot{B}_1^m \times R_+$ and Φ_2 of U_2 onto $\dot{B}_2^n \times R_+$. Let $U = U_1 \times U_2$; then $\Phi = \Phi_1 \times \Phi_2$ is a homeomorphism of U onto $\dot{B}_1^m \times \dot{B}_2^n \times R_+ \times R_+$ with the properties of ii) above. We then obtain a differentiable structure on $B_1^m \times B_2^n$ by straightening the angle.

The following will be useful in the following section.

(3.1) *Suppose P and Q are closed disjoint subsets of the compact n-manifold B^n. There exists a topological manifold $B_1^n \subset B^n$ with $P \subset B_1^n$, $B_1^n \cap Q = \emptyset$ and B_1^n closed in B^n. Moreover B_1^n can be given a differentiable structure by straightening the angle.*

Proof. By (1.2), we may as well identify a neighborhood U of $\dot{B}^n$ with $\dot{B}^n \times [0, 1)$, and we do so. By normality, there exist disjoint closed subsets P_1 and Q_1 of $\dot{B}^n$ containing $P \cap \dot{B}^n$ and $Q \cap \dot{B}^n$ respectively in their interior. By compactness there exists a $0 < t < 1$ such that

$$P_1 \times [0, t) \supset P \cap (\dot{B}^n \times [0, t)), \; Q_1 \times [0, t) \supset Q \cap (\dot{B}^n \times [0, t)) \, .$$

Let B'^n be the n-manifold $B^n \setminus \dot{B}^n \times [0, t)$, and let $P' = P \cap B'^n$, $Q' = Q \cap B'^n$. There exists a differentiable function $f' : B'^n \to [0, 1]$ with

$$f'((P_1 \times t) \cup P') = 0, \, f'((Q_1 \times t) \cup Q') = 1 \, .$$

Extend f' to a function $f : B^n \to [0, 1]$ by defining $f(x, s) = f'(x, t)$ for $(x, s) \in \dot{B}^n \times [0, t)$. Then $f(P) = 0$, $f(Q) = 1$ and f is differentiable on $\dot{B}^n \times [0, t)$. There exists a differentiable approximation F to f with $F = f$ on $\dot{B}^n \times [0, t_0]$, $0 < t_0 < t$. We may suppose that F picked so close to f that $\mathrm{l.u.b.}F(P) < \mathrm{g.l.b.}F(Q)$. There exists now an s_0 with $\mathrm{l.u.b.}F(P) < s_0 < \mathrm{g.l.b.}F(Q)$ and with s_0 a regular value of F [30]. Then $F^{-1}[0, s_0]$ is seen to be a topological n-manifold B_1^n with $B_1^n \supset P$ and $B_1^n \cap Q = \emptyset$. We leave it to the reader to see that the angle can be straightened around the $(n - 2)$-manifold $F^{-1}(s_0) \cap \dot{B}^n$ so as to give a differentiable structure to B_1^n.

4. The bordism groups of pairs of spaces

In this section we define a singular bordism theory quite analogous to singular homology.

Fix a pair (X, A) consisting of a space X and a subspace A. An *oriented singular manifold* in (X, A) is a pair (B^n, f) consisting of a compact oriented manifold B^n and a map $f : (B^n, \dot{B}^n) \to (X, A)$. If $A = \emptyset$ then of course $\dot{B}^n = \emptyset$ also.

An oriented singular manifold (B^n, f) in (X, A) is said to *bord* if an only if there is a compact oriented manifold C^{n+1} and a map $F: C^{n+1} \to X$ for which

i) B^n is contained in $\dot{C}^{n+1}$ as a regular submanifold whose orientation is induced by that of C^{n+1};

ii) $F|B^n = f$ and $f(\dot{C}^{n+1}\backslash B^n) \subset A$.

From two singular oriented manifolds (B_1^n, f_1) and (B_2^n, f_2) a disjoint union $(B_1^n \cup B_2^n, f_1 \cup f_2)$ is defined where $B_1^n \cap B_2^n = \emptyset$, $f_1 \cup f_2|B_1^n = f_1$ and $f_1 \cup f_2|B_2^n = f_2$. Define $-(B^n, f) = (-B^n, f)$. A pair (B_1^n, f_1) and (B_2^n, f_2) of oriented singular manifolds in (X, A) are *bordant* if and only if the disjoint union $(B_1^n \cup -B_2^n, f_1 \cup f_2)$ bords in (X, A). The reader may show that this defines an equivalence relation among the oriented singular manifolds in (X, A); a device similar to straightening the angle is needed in the proof of transitivity.

Denote the bordism class of (B^n, f) by $[B^n, f]$, and the collection of all such bordism classes by $\Omega_n(X, A)$. An abelian group structure is imposed on $\Omega_n(X, A)$ by disjoint union; that is,

$$[B_1^n, f_1] + [B_2^n, f_2] = [B_1^n \cup B_2^n, f_1 \cup f_2].$$

It is seen that the class of all (B^n, f) which bord forms the zero element, and that $-[B^n, f] = [-B^n, f]$. We refer to $\Omega_n(X, A)$ as the *n-dimensional oriented bordism group* of (X, A). Similar groups have been defined by ATIYAH [1], EELS, and MILNOR. In one way or another much of the remainder of this tract is devoted to the study of these groups.

Let $\Omega_*(X, A)$ be the weak direct sum $\Sigma\Omega_n(X, A)$. We define on $\Omega_*(X, A)$ the structure of a graded module over the Thom bordism ring Ω_*. From an oriented singular manifold (B^n, f) in (X, A) and a closed oriented manifold M^m, a singular manifold $(B^n \times M^m, g)$ is obtained with $g(x, y) = f(x)$. Define the module structure by

$$[B^n, f] [M^m] = [B^n \times M^m, g].$$

The reader may check that this product satisfies the appropriate associative and distributive laws.

Given a map $\varphi: (X, A) \to (X_1, A_1)$ there is associated a natural homomorphism $\varphi_*: \Omega_n(X, A) \to \Omega_n(X_1, A_1)$ given by $\varphi_*[B^n, f] = [B^n, \varphi f]$. There is also a homomorphism $\partial: \Omega_n(X, A) \to \Omega_{n-1}(A)$ given by $\partial_*[B^n, f] = [\dot{B}^n, f|\dot{B}^n]$. It is easy to see that ∂ is well defined and additive. In fact, $\varphi_*: \Omega_*(X, A) \to \Omega_*(X_1, A_1)$ and $\partial: \Omega_*(X, A) \to \Omega_*(A)$ are Ω-module homomorphisms of degree 0 and -1 respectively. We have chosen a right module structure above since $(B^n \times M^m)^{\boldsymbol{\cdot}} = \dot{B}^n \times M^m$, while $(M^m \times B^n)^{\boldsymbol{\cdot}} = (-1)^m(M^m \times \dot{B}^n)$. However, we may also introduce a left module structure and we have

$$[M^m] \, [B^n, f] = (-1)^{mn} [B^n, f] \, [M^m] \, .$$

As we have said in section 2, if $[M^m] \neq 0$ then either $m = 4k$ or $[M^m]$ is of order two. It then follows that

$$[M^m] \, [B^n, f] = [B^n, f] \, [M^m] \, .$$

5. The Eilenberg-Steenrod axioms

We have defined in section 4 a covariant functor $\{\Omega_n(X, A), \varphi_*, \partial\}$ on the category of pairs of spaces and maps of pairs. We proceed to show that it is a functor of the homology type. Such generalized homology theories have recently been studied more generally by G. M. WHITE-HEAD [43].

(5.1) **Theorem.** *On the category of pairs of spaces and maps of pairs the oriented bordism functor $\{\Omega_*(X, A), \varphi_*, \partial\}$ satisfies the first six Eilenberg-Steenrod axioms for a homology theory. However for a single point p we have $\Omega_n(p) \cong \Omega_n$, the oriented Thom bordism group.*

We proceed to enumerate the axioms; the first three are trivially verified.

(5.2) *If $i:(X, A) \to (X, A)$ is the identity map then $i_*:\Omega_n(X, A) \to \Omega_n(X, A)$ is the identity automorphism.*

(5.3) *If $\varphi:(X, A) \to (X_1, A_1)$ and $\theta:(X_1, A_1) \to (X_2, A_2)$, then $(\theta\varphi)_* = \theta_* \varphi_*$.*

(5.4) *For any map $\varphi:(X, A) \to (X_1, A_1)$ the diagram*

$$
\begin{array}{ccc}
\Omega_n(X, A) & \xrightarrow{\;\partial\;} & \Omega_{n-1}(A) \\[4pt]
\Big\downarrow{\varphi_*} & & \Big\downarrow{(\varphi|A)^*} \\[4pt]
\Omega_n(X_1, A_1) & \xrightarrow{\;\partial\;} & \Omega_{n-1}(A_1)
\end{array}
$$

is commutative.

We have next the homotopy axiom.

(5.5) *If $\varphi_0, \varphi_1:(X, A) \to (X_1, A_1)$ are homotopic, then $\varphi_{0*} = \varphi_{1*}$.*

Proof. Let $h:(I \times X, I \times A) \to (X_1, A_1)$ be a homotopy joining φ_0 and φ_1. For (B^n, f) a singular manifold in (X, A) define $\theta:I \times B^n \to X_1$ by $\theta(t, x) = h(t, f(x))$; then $\theta(0, x) = \varphi_0 f(x)$ and $\theta(1, x) = \varphi_1 f(x)$. Now $I \times B^n$ is a manifold (see section 3) and $(I \times B^n)^{\cdot} = (\dot{I} \times B^n) \cup (I \times -\dot{B}^n)$. Thus $1 \times B^n \cup 0 \times -B^n$ is a regular submanifold of the boundary, and $\theta(I \times (-\dot{B}^n)) \subset A_1$. Hence $[B^n, \varphi_0 f] = [B^n, \varphi_1 f]$.

We need a remark before proving exactness. Let $V^n \subset M^n$ be a regular submanifold with boundary in a closed manifold M^n. If $f:M^n \to X$ is a map with $f(M^n \backslash \operatorname{Int} V^n) \subset A$, then $[M^n, f] = [V^n, f|V^n]$ in $\Omega_n(X, A)$. Define $F:I \times M^n \to X$ by $F(t, x) = f(x)$. Now $(I \times M^n)^{\cdot} = \dot{I} \times M^n$, and

$1 \times V^n \cup 0 \times - M^n$ is a regular submanifold of the boundary. Since $F(1 \times (M^n\backslash\mathrm{Int}\,V^n)) \subset A$ then $[M^n, f] = [V^n, f\,|\,V^n]$ in $\Omega_n(X, A)$.

(5.6) *For every pair* (X, A) *the sequence*

$$\cdots \longrightarrow \Omega_n(A) \xrightarrow{i_*} \Omega_n(X) \xrightarrow{j_*} \Omega_n(X, A) \xrightarrow{\partial} \Omega_{n-1}(A) \longrightarrow \cdots$$

is exact.

Proof. It is easy to see that $\partial j_* = 0$ and $i_* \partial = 0$. Consider now $[M^n, f] \in \Omega_n(A)$. We can apply the preceding remark to see that $j_* i_* [M^n, f] = 0$ in $\Omega_n(X, A)$. Next consider $[C^n, f] \in \Omega_n(X, A)$ in the kernel of ∂. There is a B^n and a map $g: B^n \to A$ with $\dot{B}^n = C^n$ and $g\,|\,\dot{B}^n = f\,|\,\dot{C}^n$. Identify C^n and $-B^n$ along their common boundary to obtain closed oriented manifold M^n and a map $F: M^n \to X$ with $F\,|\,C^n = f$ and $F\,|\,B^n = g$. Now $[M^n, F] \in \Omega_n(X)$; that $j_* [M^n, F] = [C^n, f]$ in $\Omega_n(X, A)$ follows from the remark preceding (5.6). The remainder of exactness is trivial.

(5.7) *If* $\overline{U} \subset \mathrm{Int}\,A$, *then the inclusion* $i: (X\backslash U, A\backslash U) \subset (X, A)$ *induces an isomorphism*

$$i_*: \Omega_n(X\backslash U, A\backslash U) \approx \Omega_n(X, A)\,.$$

Proof. We show that i_* is an epimorphism; the remainder of the argument is similar. Let (B^n, f) be an oriented singular manifold in (X, A). Let $P = f^{-1}(X\backslash\mathrm{Int}\,A)$, $Q = f^{-1}(\overline{U})$. There exists, by (3.1), a topological submanifold $B_1^n \subset B^n$ with $B_1^n \supset P$ and $B_1 \cap Q = \emptyset$. Furthermore B_1^n can be given a differentiable structure by straightening the angle.

Now $[B_1^n, f\,|\,B_1^n] \in \Omega_n(X\backslash U, A\backslash U)$. That $i_* [B_1^n, f\,|\,B_1^n] = [B^n, f]$ in $\Omega_n(X, A)$ follows just as in the remark preceding (5.6).

6. Consequences of the axioms

A number of facts follow from the first six Eilenberg-Steenrod axioms alone; here the general reference is EILENBERG-STEENROD [19]. For one thing we now have reduced bordism groups. Identify the Thom group Ω_n with $\Omega_n(p)$ for any point p. Define the reduced group $\tilde{\Omega}_n(X)$ to be the kernel of $\varepsilon_*: \Omega_n(X) \to \Omega_n(p)$ where ε collapses X to the point p. As usual $\Omega_n(X) \approx \Omega_n \oplus \tilde{\Omega}_n(X)$. It is seen that an oriented singular manifold (M^n, f) in X represents an element of $\tilde{\Omega}_n(X)$ if and only if $[M^n] = 0$ in Ω_n. For each pair (X, A) there is an exact sequence [19, p. 20]

$$\cdots \to \tilde{\Omega}_n(A) \to \tilde{\Omega}_n(X) \to \Omega_n(X, A) \to \tilde{\Omega}_{n-1}(A) \to \cdots.$$

Also for every triple $X \supset A \supset B$ there is the exact sequence [19, p. 25]

$$\cdots \to \Omega_n(A, B) \to \Omega_n(X, B) \to \Omega_n(X, A) \to \cdots.$$

There are also a number of consequences for pairs (X, A) consisting of a CW complex X and a closed subcomplex A; these are proved just as in EILENBERG-STEENROD [19]. For example, full excision holds in this category. Also if $\varphi: (X, A) \to (X_1, A_1)$ is a relative homeomorphism between CW pairs, then $\varphi_*: \Omega_n(X, A) \cong \Omega_n(X_1, A_1)$. In particular, $\Omega_n(X, A) \cong \tilde{\Omega}_n(X/A)$. Here if $A = \emptyset$ we agree by convention that $X/\emptyset$ is the disjoint union of X and a point. There is also a Mayer-Vietoris sequence and a relative Mayer-Vietoris sequence for CW triads [19, p. 76].

One can now compute such groups as

$$\tilde{\Omega}_n(S^k) \cong \Omega_{n-k}, \quad \Omega_n(S^k) \cong \Omega_n \oplus \Omega_{n-k},$$

and

$$\Omega_n(I^k, S^{k-1}) \cong \tilde{\Omega}_{n-1}(S^{k-1}) \cong \Omega_{n-k}.$$

For pairs (X, A) and (Y, B) there is a homomorphism

$$\varkappa: \Omega_p(X, A) \otimes \Omega_q(Y, B) \to \Omega_{p+q}((X, A) \times (Y, B))$$

where $(X, A) \times (Y, B) = (X \times Y, A \times Y \cup X \times B)$. Namely, define

$$\varkappa([B^p, f] \otimes [C^q, g]) = [B^p \times C^q, f \times g],$$

where $B^p \times C^q$ has the product orientation. If Y is a point and B is empty, we obtain

$$\varkappa: \Omega_p(X, A) \otimes \Omega_q \to \Omega_{p+q}(X, A),$$

which is just the Ω-module structure on $\Omega_*(X, A)$ already described.

There is also a natural homomorphism

$\mu: \Omega_n(X, A) \to H_n(X, A; Z)$, where $H_n(X, A; Z)$ is the integral singular homology group. Given $[B^n, f] \in \Omega_n(X, A)$, let $\sigma_n \in H_n(B^n, \dot{B}^n; Z)$ denote the orientation class of B^n. Define $\mu[B^n, f]$ to be the element $f_*(\sigma_n) \in H_n(X, A; Z)$. The image of μ is the subgroup of integral homology classes representable in the sense of STEENROD. It will be recalled that STEENROD raised the following question [18, p. 257]: given an integral homology class γ of a complex X, is there a map of a closed oriented manifold into X carrying the orientation class into γ?

We note that commutativity holds in

$$(6.1) \quad \begin{array}{ccc} \Omega_n(X, A) \xrightarrow{\;\mu\;} H_n(X, A) & \quad & \Omega_n(X, A) \xrightarrow{\;\mu\;} H_n(X, A) \\ \downarrow{\varphi_*} \qquad\qquad \downarrow{\varphi_*} & \quad & \downarrow{\partial} \qquad\qquad \downarrow{\partial} \\ \Omega_n(X_1, A_1) \xrightarrow{\;\mu\;} H_n(X_1, A_1) & \quad & \Omega_{n-1}(A) \xrightarrow{\;\mu\;} H_{n-1}(A). \end{array}$$

As in EILENBERG-STEENROD [19, p. 45], there are isomorphisms

$$\Omega_r(I^n, S^{n-1}) \cong \Omega_{r-1}(I^{n-1}, S^{n-2}),$$

the composition of a boundary homomorphism and homomorphisms induced by maps. We then have a commutative diagram

$$\begin{array}{ccc}
\Omega_n(I^n, S^{n-1}) & \xrightarrow{\mu} & H_n(I^n, S^{n-1}) \\
\downarrow & & \downarrow \\
\Omega_{n-1}(I^{n-1}, S^{n-1}) & \xrightarrow{\mu} & H_{n-1}(I^{n-1}, S^{n-1}) .
\end{array}$$

It follows by induction that $\mu : \Omega_n(I^n, S^{n-1}) \cong H_n(I^n, S^{n-1})$, since this is the case for $n = 0$. By the relative homeomorphism property and the direct sum theorems it follows for any CW complex X that

$$\mu : \Omega_n(X^n, X^{n-1}) \cong H_n(X^n, X^{n-1}) ,$$

where X^k denotes the k-skeleton. Moreover commutativity holds in

$$\begin{array}{ccccc}
\Omega_n(X^n, X^{n-1}) & \cong & H_n(X^n, X^{n-1}) & = & C_n(X) \\
\downarrow \partial & & \downarrow \partial & & \downarrow \partial \\
\Omega_{n-1}(X^{n-1}, X^{n-2}) & \cong & H_{n-1}(X^{n-1}, X^{n-2}) & = & C_{n-1}(X) .
\end{array}$$

(6.2)

We note next that

(6.3) $$\varkappa : \Omega_n(I^n, S^{n-1}) \otimes \Omega_q \cong \Omega_{n+q}(I^n, S^{n-1}) .$$

For first of all (6.3) holds for $n = 0$. The general case follows by induction on n from

$$\begin{array}{ccc}
\Omega_n(I^n, S^{n-1}) \otimes \Omega_q & \to & \Omega_{n+q}(I^n, S^{n-1}) \\
\downarrow & & \downarrow \\
\Omega_{n-1}(I^{n-1}, S^{n-2}) \otimes \Omega_q & \to & \Omega_{n+q-1}(I^{n-1}, S^{n-2})
\end{array}$$

It then follows that

(6.4) $$\varkappa : \Omega_n(X^n, X^{n-1}) \otimes \Omega_q \cong \Omega_{n+q}(X^n, X^{n-1}) .$$

Denote by SX the usual suspension of X, namely $X \times I$ with $X \times 0$ and $X \times 1$ each identified to a point. Identify X with $X \times {}^1/_2$; there are then the cones

$$C_- = X \times [0, {}^1/_2]/X \times 0, \quad C_+ = X \times [{}^1/_2, 1]/X \times 1$$

in SX, and the composite isomorphism

$$\Delta : \tilde{\Omega}_n(SX) \cong \Omega_n(SX, C_+) \cong \Omega_n(C_-, X) \cong \tilde{\Omega}_{n-1}(X) .$$

There are also maps $\varphi : X/A \to SA$ of the type well-known in stable homotopy. In our case we extend the identity map $A \to A$ to a map $(X, A) \to (C_-, A)$, for (X, A) a CW pair, and then define φ as the composite

$$X/A \to C_-/A \cong SA .$$

(6.5) *The composite* $\Omega_n(X, A) \cong \tilde{\Omega}_n(X/A) \xrightarrow{\varphi_*} \tilde{\Omega}_n(SA) \xrightarrow{\Delta} \tilde{\Omega}_{n-1}(A)$ *is the homomorphism* $\partial : \Omega_n(X, A) \to \tilde{\Omega}_{n-1}(A)$.

This follows immediately from the commutative diagram

$$
\begin{array}{ccccc}
\tilde{\Omega}_n(X/A) & \xleftarrow{\cong} & \Omega_n(X,A) & \xrightarrow{\ \partial\ } & \tilde{\Omega}_{n-1}(A) \\
\ \downarrow{\varphi_*} & & \ \downarrow{\varphi_*} & & \ \downarrow{=} \\
\tilde{\Omega}_n(SA) & \xleftarrow{\cong} & \Omega_n(C_-,A) & \longrightarrow & \tilde{\Omega}_{n-1}(A)\ .
\end{array}
$$

We can also define a suspension homomorphism $S:\tilde{\Omega}_{n-1}(X) \to$ $\to \tilde{\Omega}_n(SX)$. Suppose $[M^{n-1}, f]$ represents an element of $\tilde{\Omega}_{n-1}(X)$. Then $M^{n-1} = \dot{B}^n$ for some compact manifold B^n. There is a map F of $(B^n \times I)^{\cdot}$ into S which maps $M^{n-1} \times I$ into $A \times I$ via $f \times id$, and which maps $B^n \times 1$ into the north pole and $B^n \times 0$ into the south pole. The element $[(B^n \times I)^{\cdot}, F]$ is uniquely determined, since

(6.6) $\Delta [(B^n \times I)^{\cdot}, F] = (-1)^{n-1}[M^{n-1}, f]$. Define $S[M^{n-1}, f]$ to be $[(B^n \times I)^{\cdot}, F]$ in $\tilde{\Omega}_n(X)$.

7. The bordism spectral sequence

For a CW pair (X, A) there are the groups $\Omega_p(X^k \cup A, X^l \cup A)$, X^k the k-skeleton of X, and the exact sequences of the triples

$$X^k \cup A \supset X^l \cup A \supset X^m \cup A, \quad k \geqq l \geqq m\ .$$

We thus receive a spectral sequence, whose elementary properties we study in this section.

Using the terminology of EILENBERG [11, Chapter 8], there is a spectral sequence $\{E^r_{p,q}\}$ where $E^r_{p,q} = C^r_{p,q}/B^r_{p,q}$, and where $C^r_{p,q}$ and $B^r_{p,q}$ are the images of

$$\Omega_{p+q}(X^p \cup A, X^{p-r} \cup A) \to \Omega_{p+q}(X^p \cup A, X^{p-1} \cup A)\ ,$$

$$\Omega_{p+q+1}(X^{p+r-1} \cup A, X^p \cup A) \to \Omega_{p+q}(X^p \cup A, X^{p-1} \cup A)\ .$$

Moreover there is the filtration

$$0 \subset J_{0,n} \subset \cdots \subset J_{p,n-p} \subset \cdots \subset J_{n,0} = \Omega_n(X, A)$$

with $J_{p,q}$ the image of

$$\Omega_{p+q}(X^p \cup A, A) \to \Omega_{p+q}(X, A)$$

and $J_{p,q}/J_{p-1,q+1} \cong E^\infty_{p,q}$.

We have $E^1_{p,q} \cong \Omega_{p+q}(X^p \cup A, X^{p-1} \cup A) \cong C_p(X, A) \otimes \Omega_q \cong$ $\cong C_p(X, A; \Omega_q)$ by (6.4) and (6.2). Moreover $d^1_{p,q}:E^1_{p,q} \to E^1_{p-1,q}$ is identified with $\partial:C_p(X,A;\Omega_q) \to C_{p-1}(X,A;\Omega_q)$. Hence $E^2_{p,q} = H_p(X,A;\Omega_q)$.

We have thus for each CW pair a spectral sequence $\{E^r_{p,q}, d^r\}$, with E^∞-term associated with a filtration of $\Omega_*(X, A)$. If $\varphi:(X, A) \to (X_1, A_1)$ is cellular there is induced a homomorphism of the spectral sequence of (X, A) into that of (X_1, A_1), compatible with $\varphi_*: \Omega_*(X, A) \to \Omega_*(X_1, A_1)$. If $\varphi_0, \varphi_1:(X, A) \to (X_1, A_1)$ are cellularly homotopic it follows in a standard way that the homomorphisms $E^r_{p,q} \to E'^r_{p,q}$ agree for $r \geqq 2$.

If $\varphi:(X, A) \to (X_1, A_1)$ is a map joining CW pairs, there is a homotopic map φ' which is cellular [44, p. 221]. Define $\varphi_*: E^r_{p,q} \to E'^r_{p,q}$ to be φ'_*. This is well-defined for $r \geqq 2$ since homotopic cellular maps are cellularly homotopic [44]. In particular the bordism spectral sequence is independent of the particular cell subdivision for $r \geqq 2$.

Note also that the Thom ring Ω acts on the spectral sequence. For each $\Omega_*(X^k \cup A, X^l \cup A)$ is an Ω-module, and the inclusion and boundary homomorphisms are Ω-module homomorphisms. Hence $E^n = \Sigma E^n_{p,q}$ is an Ω-module, each d^n is an Ω-module homomorphism and $E^{n+1} \cong H(E^n)$ is an Ω-isomorphism. Finally the product $\varkappa: \Omega_*(X, A) \otimes \Omega \to \Omega_*(X, A)$ has $\varkappa(J_{p,q} \otimes \Omega_s) \subset J_{p,q+s}$. The induced

$$\varkappa':(J_{p,q}/J_{p-1,q+1}) \otimes \Omega_s \to J_{p,q+s}/J_{p-1,q+s+1}$$

is the homomorphism $E^\infty_{p,q} \otimes \Omega_s \to E^\infty_{p,q+s}$.

(7.1) *The product $E^2_{p,q} \otimes \Omega_s \to E^2_{p,q+s}$ of the spectral sequence is identified with the homomorphism*

$$H_p(X, A; \Omega_q) \otimes \Omega_s \to H_p(X, A; \Omega_q \otimes \Omega_s) \to H_p(X, A; \Omega_{q+s}).$$

Proof. The remark follows from the commutative diagram

$$
\begin{array}{ccc}
E^1_{p,q} \otimes \Omega_s & \longrightarrow & E^1_{p,q+s} \\
\Big\downarrow{\cong} & & \Big\downarrow{\cong} \\
C_p(X, A; \Omega_q) \otimes \Omega_s \to C_p(X, A; \Omega_q \otimes \Omega_s) \to C_p(X, A; \Omega_{q+s}).
\end{array}
$$

(7.2) *The edge homomorphism $\Omega_n(X, A) = J_{n,0} \to E^\infty_{n,0} \to E^2_{n,0} = H_n(X, A)$ of the bordism spectral sequence coincides with the homomorphism $\mu: \Omega_n(X, A) \to H_n(X, A)$ of* (6.1).

Proof. The edge homomorphism may be described as follows. If $b \in \Omega_n(X, A) = J_{n,0}$ then b is the image of an element c under $\Omega_n(X^n \cup A, A) \to \Omega_n(X, A)$. There is the image c' of c under $\Omega_n(X^n \cup A, A) \to \Omega_n(X^n \cup A, X^{n-1} \cup A)$. Then $c' \in E^1_{n,0} = C_n(X, A)$, and c' represents an element $d \in E^2_{n,0}$. The edge homomorphism maps b into d.

We now prove (7.2). Consider an element $[B^n, f] \in \Omega_n(X, A)$. There is $[B^n, id] \in \Omega_n(B^n, \dot{B}^n)$ and $f_*[B^n, id] = [B^n, f]$. Naturality shows that it suffices to prove (7.2) for $[B^n, id] \in \Omega_n(B^n, \dot{B}^n)$. Under the edge homomorphism we must show that $[B^n, id]$ maps into the orientation class of B^n. Letting $b = [B^n, id] \in \Omega_n(B^n, \dot{B}^n)$, we have $c = b$. Hence c' is

$[B^n, id] \in \Omega_n(B^n, X^{n-1})$, where X^{n-1} is the $(n-1)$-skeleton of B^n. Under the identification

$$\mu: \Omega_n(B^n, X^{n-1}) \cong H_n(B^n, X^{n-1}) = C_n(B^n) ,$$

c' is identified with the orientation cycle by the definition of μ. The theorem follows.

8. Unoriented bordism groups

We sketch here the unoriented version $\mathfrak{N}_n(X, A)$ of the oriented bordism group $\Omega_n(X, A)$. As one would guess from the work of THOM, the structure of $\mathfrak{N}_n(X, A)$ is easier to determine than that of $\Omega_n(X, A)$; in fact, it is not difficult to determine it completely.

A *singular manifold* in (X, A) is a map $f:(B^n, \dot{B}^n) \to (X, A)$, where B^n is a compact manifold. A bordism relation is defined just as in § 4, except that no orientability requirements are imposed. The resulting group of bordism classes is denoted by $\mathfrak{N}_n(X, A)$; every element is of order two. The bordism class of (B^n, f) is denoted by $[B^n, f]_2$. The weak direct sum $\mathfrak{N}_*(X, A) = \Sigma \mathfrak{N}_n(X, A)$ is a graded $\mathfrak{N}_*$-module. The functor $\{\mathfrak{N}_n(X, A), \varphi_*, \partial\}$ is defined on the category of pairs of spaces and maps of pairs. This unoriented bordism functor satisfies the first six Eilenberg-Steenrod axioms for a homology theory; however for a single point p we have $\mathfrak{N}_n(p) \cong \mathfrak{N}_n$, the unoriented Thom group.

There is a natural homomorphism $\mu:\mathfrak{N}_n(X, A) \to H_n(X, A; Z_2)$ defined just as in § 6. Moreover for every CW pair (X, A) there is an unoriented bordism spectral sequence $\{E^r_{p,q}, d^r\}$ with $E^2_{p,q} \cong H_p(X, A; \mathfrak{N}_q)$ and whose E^∞-term is associated with a filtration of $\mathfrak{N}_*(X, A)$.

In the unoriented case we see that $H_p(X, A; \mathfrak{N}_q) \cong H_p(X, A; Z_2) \otimes \mathfrak{N}_q$, and in fact $\varkappa: E^2_{p,0} \otimes \mathfrak{N}_q \cong E^2_{p,q}$ for all p, q.

(8.1) **Thom.** *For each CW pair (X, A), $\mu:\mathfrak{N}_n(X, A) \to H_n(X, A; Z_2)$ is an epimorphism.*

This was shown by THOM in [40], and we shall now discuss its implications. In the unoriented bordism spectral sequence of (X, A), it follows from the unoriented version of (7.2) that $E^2_{p,0}$ consists entirely of permanent cycles. Then $d^2_{p,0} = 0$; since $E^2_{p,q} \cong E^2_{p,0} \otimes \Omega_q$, it follows that $d^2_{p,q} = 0$ and $E^3_{p,q} = E^2_{p,q}$. Continuing in the same fashion, we see that $d^r_{p,q} = 0$ for $r \geq 2$. We have the following corollary.

(8.2) **Theorem.** *For every CW pair (X, A), the unoriented bordism spectral sequence is trivial.*

If $\mathfrak{N}_n(X, A) = J_{n,0} \supset \cdots \supset J_{0,n}$ is the filtration coming from the spectral sequence, then $J_{n-q, q}/J_{n-q-1, q+1} \cong E^\infty_{n-q, q} \cong E^2_{n-q, q} \cong H_{n-q}(X, A; \mathfrak{N}_q)$.

Since every element in $\mathfrak{N}_n(X, A)$ has order two, we thus see that $\mathfrak{N}_n(X, A) \cong \Sigma H_{n-q}(X, A; Z_2) \otimes \mathfrak{N}_q$. A refinement of this last remark is possible.

(8.3) Theorem. *For every CW pair* (X, A), $\mathfrak{N}_*(X, A)$ *is a free graded* $\mathfrak{N}$-*module isomorphic to* $H_*(X, A; Z_2) \otimes \mathfrak{N}$.

Proof. We do not prove this here in complete detail; that will be done in Section 17. Let $\{c_{n,i}\} \subset H_*(X, A; Z_2)$ be an additive homogeneous base. For each $c_{n,i}$ select an unoriented singular manifold (B_i^n, f_i) in (X, A) with $\mu([B_i^n, f_i]_2) = c_{n,i}$. The elements $\{[B_n^n, f_i]_2\}$ can be shown to form a homogeneous $\mathfrak{N}$-base of $\mathfrak{N}_*(X, A)$.

It is easily seen from (8.2) that $\{[B_i^n, f_i]_2\}$ forms a generating set for $\mathfrak{N}_*(X, A)$. We shall later in passing show the independence of the base.

9. Differentiable bordism groups

Let X^k be a differentiable manifold without boundary, with no requirements of compactness or orientability. We define differentiable bordism groups $D_n(X^k)$.

To do so, consider pairs (M^n, f) consisting of a closed oriented manifold M^n and a differentiable map $f: M^n \to X^k$. Such a pair orientably bords if and only if there is a compact oriented manifold B^{n+1} with $\dot{B}^{n+1} = M^n$ and a differentiable map $g: B^{n+1} \to X^k$, with $g \mid \dot{B}^{n+1} = f$, and such that there exist an open set U containing $\dot{B}^{n+1}$ and a diffeomorphism $h: \dot{B}^{n+1} \times [0, 1) \to U$ with $h(x, 0) = x$ and with $g(h(x, t)) = f(x)$ for all $0 \leq t < 1$ and $x \in \dot{B}^{n+1}$. This last condition is seen to guarantee the transitivity of the bordism relation. The resulting group of bordism classes is denoted by $D_n(X^k)$.

There is a natural homomorphism $D_n(X^k) \to \Omega_n(X^k)$, mapping the class of (M^n, f) into the class of (M^n, f). This is an epimorphism, for given $f: M^n \to X^k$ there is for $\varepsilon > 0$ an ε-approximation $f': M^n \to X^k$ with f' differentiable. Since f and f' are homotopic for ε sufficiently small, $[M^n, f] = [M^n, f']$ in $\Omega_n(X^k)$.

We next show that $D_n(X^k) \to \Omega_n(X^k)$ is a monomorphism. For let $f: M^n \to X^k$ be differentiable, and suppose there is a compact oriented manifold B^{n+1} with $\dot{B}^{n+1} = M^n$ and a map $g: B^{n+1} \to X^k$ with $g \mid \dot{B}^{n+1} = f$. Choose a neighborhood U of $\dot{B}^{n+1}$ in B^{n+1} which is diffeomorphic to $\dot{B}^{n+1} \times [0, 1)$, and identify the two. The map $g: \dot{B}^{n+1} \times [0, 1/2] \to X^k$ is seen to be homotopic to the map $g': \dot{B}^{n+1} \times [0, 1/2] \to X^k$ given by $g'(x, t) = f(x)$. By the homotopy extension property, there is then a map $g': B^{n+1} \to X^k$ with $g'(x, t) = f(x)$ for $0 \leq t \leq 1/2$. By the approximation theorems for differentiable maps [22], there is an ε-approximation $g'': B^{n+1} \to X^k$ with g'' differentiable and $g'' = g'$ on $\dot{B}^{n+1} \times [0, 1/3]$. Thus f differentiably bords, and we receive the following theorem.

(9.1) Theorem. *If* X^k *is a differentiable manifold without boundary, then* $D_n(X^k) \cong \Omega_n(X^k)$.

Similar remarks apply to the unoriented bordism groups.

10. A review of differential topology

Up to this point we have made do with the very simplest differential topology. However the situation soon becomes more complicated. Hence we give at this stage a summary of some of the things assumed.

First there are some purely geometric facts; here MILNOR [22] is the appropriate general reference. First of all, there is the fact that every continuous function can be approximated by a differentiable function. We state the following (see MILNOR [22, p. 62]).

(10.1) *Let $f: M \to N$ be a continuous map of differentiable manifolds without boundary, with f differentiable on the closed subset A of M. Let a positive real. — valued function ε be given on M, and let N have the metric determined by an embedding $N \subset R^p$. Then there exists $g: M \to N$ with g differentiable, g an ε-approximation of f, and with $g|A = f|A$.*

For a differentiable manifold M without boundary, let M_x denote the vector space of tangent vectors to M at $x \in M$. If $f: M \to N$ is differentiable, for each $x \in M$ there is the homomorphism $df: M_x \to N_{f(x)}$, the differential of f.

The map f is an *immersion* if df is a monomorphism for each $x \in M$, and an *embedding* if f is an immersion and also a homeomorphism of M onto $f(M)$. There is now the Whitney embedding theorem (see MILNOR [22, p. 21]).

(10.2) Whitney embedding theorem.
If $p > 2n$, any map f of the differentiable manifold M^n, without boundary, into R^p can be ε-approximated by an embedding g. If f is already an embedding on a neighborhood of the closed set $A \subset M^n$, we may choose $g|A = f|A$.

It is easy to generalize (10.2), replacing R^p by an arbitrary N^p without boundary.

We shall also be considering manifolds B^n with boundary. Denote by U a neighborhood of $\dot{B}^n$ which is diffeomorphic to $\dot{B}^n \times [0, 1)$; identify U with $\dot{B}^n \times [0,1)$. A map h of B^n into the solid p-ball C^p is an *embedding* if h is a $1 - 1$ immersion with $h(\dot{B}^n) \subset S^{p-1}$, $h(B^n) \cap S^{p-1} = h(\dot{B}^n)$, and if there exists $0 < t_0 < 1$ with $h(x, t) = (1 - t)\, h(x)$ for $(x, t) \in \dot{B}^n \times [0, t_0)$.

(10.3) *If B^{n+1} is a compact $(n + 1)$-manifold and if $p \geqq 2n + 2$, then every embedding of $\dot{B}^{n+1}$ into S^p can be extended to an embedding of B^{n+1} into the solid $(p + 1)$-ball C^{p+1}.*

Proof. Let U be a neighborhood of $\dot{B}^{n+1}$ diffeomorphic to $\dot{B}^{n+1} \times [0, 1)$, and identify the two. Let $h: \dot{B}^{n+1} \to S^p$ be an embedding. Define $h': B^{n+1} \to C^{p+1}$ by $h'(x, t) = (1 - t)\, h(x)$ for $(x, t) \in \dot{B}^{n+1} \times [0, 1)$, $h' = 0$ otherwise. One now uses the Whitney embedding theorem on the manifold $B^{n+1} \setminus \dot{B}^{n+1} \times [0, 1/3]$, requiring that the approximation be an extension of h' on $\dot{B}^{n+1} \times (1/3, 2/3]$. It is seen that we thus get an embedding of B^{n+1} in C^{p+1}.

We turn now to Thom's concept of transverse regularity [40, 22], invented to study such topics as cobordism. Suppose that N is a differentiable manifold without boundary, and that N' is a regularly embedded submanifold of N. The tangent space N'_x can be regarded as a subspace of the tangent space N_x for each $x \in N'$. The space of normal vectors to N' is by definition the vector space N_x/N'_x.

Suppose that M and N are differentiable manifolds without boundary, and that $f: M \to N$ is differentiable. If N' is a regularly embedded submanifold of N, then f is *transverse regular* on N' if for each $x \in f^{-1}(N')$ the composite map

$$M_x \xrightarrow{\;df\;} N_x \to N_x/N'_x$$

is an epimorphism. It is then the case that $f^{-1}(N')$ is a regularly embedded submanifold of M, and that $\dim M - \dim f^{-1}(N') = \dim N - \dim N'$.

We have now the basic approximation theorem of Thom (see Milnor [22, p. 22]).

(10.4) *Let $f: M^n \to N^p$ be differentiable, and let N_1^{p-q} be a closed differentiable submanifold of N. Let A be a closed subset of M such that the transverse regularity condition for f and N_1 holds at each x in $A \cap f^{-1}(N_1)$. Let δ be a positive real-valued continuous function on M. There exists a differentiable map $g: M^n \to N^p$ such that*

    (1) *g is a δ-approximation of f,*
    (2) *g is transverse regular on N_1, and*
    (3) *$g|A = f|A$.*

We shall also need the existence of tubular neighborhoods. Let M^n be a closed differentiable manifold. There is a Riemannian metric on M^n [38, p. 58], which we assume fixed once and for all. The tangent bundle to M^n thus receives a continuous inner product. Let V^m be a closed differentiable submanifold of M^n. The bundle $\tau: E \to V^m$, induced on V^m by the tangent bundle to M^n, splits as a Whitney sum $\tau = \tau_1 \oplus \tau_2$ where $\tau_1: E_1 \to V^m$ is the tangent bundle to V^m and $\tau_2: E_2 \to V^m$ is the orthogonal complement of E_1 in E. Moreover, τ_2 is isomorphic to the normal bundle to V^m in M^n; we identify the two.

We define now a map $h: E_2 \to M^n$, following a classical procedure. A normal vector v at $x \in V^m$ has a length $\|v\|$. If $\|v\| > 0$, there is a unique geodesic $u(s)$ in M^n, parameterized by arc length, with $u(0) = x$ and with initial direction $v/\|v\|$. The map $h: E_2 \to M^n$ is defined by $h(v) = u(\|v\|)$ if $v \neq 0$, $h(v) = x$ if $v = 0$. It is a standard fact that the Jacobian of h is non-singular along $V^m \subset E_2$, where V^m is identified with the trivial cross-section of E_2. The map $h|V^m$ being a diffeomorphism of the compact space V^m and h having non-singular Jacobian on V^m, there is an open set W, $V^m \subset W \subset E_2$, such that $h: W \to M^n$ is a diffeomorphism onto an open subset of M^n containing V^m.

Select $\varepsilon > 0$ so that if $\|v\| \leq \varepsilon$ then $v \in W$. Let $\gamma: A \to V^m$ denote the closed unit cell bundle in E_2; i.e., $A = \{v : v \in E_2$ and $\|v\| \leq 1\}$. There is then a diffeomorphism $h': A \to M^n$ onto a submanifold in M^n, given by $h'(v) = h(\varepsilon v)$. Call the image $h'(A)$ a *tubular neighborhood* of V^m of radius ε.

We turn now to a very quick resume of the theory of characteristic classes following BOREL [4, 5]. For expositions of characteristic classes, see also HIRZEBRUCH [20], MILNOR [24], or BOREL-HIRZEBRUCH [7]. For G a compact Lie group, a *universal G-bundle* is a principal G-bundle $\tau: E(G) \to B(G)$ with $E(G)$ pathwise connected and with $\pi_i(E(G)) = 0$ for $0 \leq i < \infty$, and with $B(G)$ a CW complex. Call $B(G)$ a *classifying space* for G. If $B(G)$ and $B'(G)$ are two such classifying spaces, $H^*(BG; K)$ and $H^*(B'G; K)$ are canonically isomorphic. For every inclusion $G_1 \subset G_2$ there is defined a homomorphism

$$\varrho = \varrho(G_1, G_2) : H^*(BG_2; K) \to H^*(BG_1; K) \,.$$

For the cyclic group Z_2, a classifying space BZ_2 is seen to be infinite dimensional real projective space P_∞, the CW complex which is the union of an ascending union $P_1 \subset P_2 \subset \cdots$ of real projective spaces. Hence

$$H^*(BZ_2; Z_2) = Z_2[t] \,,$$

a polynomial algebra with a single one-dimensional generator t. Generally it is the case that for $B(G_1 \times G_2)$ we may take $B(G_1 \times G_2) = BG_1 \times BG_2$. Hence

$$H^*(B(Z_2)^n; Z_2) = Z_2[t_1, \ldots, t_n] \,,$$

with each t_i one-dimensional.

Pass now to the orthogonal group $O(n)$. There is the subgroup D of diagonal orthogonal matrices in $O(n)$, and $D \cong (Z_2)^n$. BOREL [4] has shown that the inclusion $(Z_2)^n \subset O(n)$ induces a monomorphism

$$\varrho : H^*(BO(n); Z_2) \to H^*(B(Z_2)^n; Z_2)$$

whose image is the symmetric polynomials in $t_1, \ldots, t_n$. In particular there are the elementary symmetric polynomials $\Sigma_{i_1 < \cdots < i_k} t_{i_1} \cdots t_{i_k}$, denoted simply by $\Sigma t_1 \ldots t_k$. Let $w_k \in H^k(BO(n); Z_2)$ be defined by $\varrho(w_k) = \Sigma t_1 \ldots t_k$. The w_k, $1 \leq k \leq n$, are the *universal Whitney classes* and $w = 1 + w_1 + \cdots + w_n$ is the total Whitney class. Also

$$H^*(BO(n); Z_2) = Z_2[w_1, \ldots, w_n] \,.$$

There is the inclusion $SO(n) \subset O(n)$, inducing $\varrho : H^*(BO(n); Z_2) \to H^*(BSO(n); Z_2)$. It is known [4] that ϱ is an epimorphism and that its kernel is the ideal generated by w_1. Letting $\bar{w}_k = \varrho(w_k)$, $2 \leq k \leq n$, then

$$H^*(BSO(n); Z_2) = Z_2[\bar{w}_2, \ldots, \bar{w}_n] \, .$$

For the circle group S^1, a classifying space BS^1 is the infinite dimensional complex projective space $P_\infty(C)$. Hence

$$H^*(BS^1; Z) = Z[y] \, ,$$

where y is two dimensional. Moreover

$$H^*(B(S^1)^n; Z) = Z[y_1, \ldots, y_n].$$

Consider now the unitary group $U(n)$. There is the subgroup D of diagonal unitary matrices in $U(n)$, and $D \cong (S^1)^n$. BOREL [8] has shown that $(S^1)^n \subset U(n)$ induces a monomorphism $H^*(BU(n);Z) \to H^*(B(S^1)^n;Z)$ whose image is the group of symmetric polynomials in $y_1, \ldots, y_n$. Define the universal Chern class $c_k \in H^{2k}(BU(n);Z)$, $1 \leq k \leq n$, by $\varrho(c_k) = \Sigma y_1 \ldots y_k$; that is, $\varrho(1 + c_1 + \cdots + c_n) = \Pi(1 + y_k)$. Then

$$H^*(BU(n); Z) = Z[c_1, \ldots, c_n] \, .$$

Consider again $O(2n)$. There is the embedding of $(S^1)^n$ in $O(2n)$, with $(z_1, \ldots, z_n) \in (S^1)^n$, $z_k = x_k + iy_k$, identified with the block matrix $\begin{pmatrix} \alpha_1 & & 0 \\ & \ddots & \\ 0 & & \alpha_k \end{pmatrix}$ where

$$\alpha_k = \begin{pmatrix} x_k & -y_k \\ y_k & x_k \end{pmatrix}$$

The induced homomorphism $\varrho: H^*(BO(2n); Z) \to H^*(B(S^1)^n; Z)$ maps $H^*(BO(2n); Z)$ onto the symmetric polynomials $S[y_1^2, \ldots, y_n^2]$. Moreover the kernel of ϱ is the 2-torsion of $H^*(BO(2n); Z)$, which consists solely of elements of order two; for all this, see BOREL-HIRZEBRUCH [7]. There are the inclusions

$$(S^1)^n \subset U(n) \subset O(2n) \subset U(2n) \, ,$$

and $\varrho: H^*(BU(2n); Z) \to H^*(B(S^1)^n; Z)$ maps $(1 + c_1 + c_2 + \cdots)(1 - c_1 + c_2 - \cdots)$ into $\Pi(1 - y_k^2)$. Define the universal Pontryagin classes $p_k \in H^{4k}(BO(2n); Z)$, $1 \leq k \leq n$, by $1 - p_1 + p_2 - \cdots + (-1)^n p_n = \varrho[(1 + c_1 + \cdots + c_n)(1 - c_1 + \cdots + (-1)^n c_n)]$ where ϱ is the homomorphism $H^*(BU(2n)) \to H^*(BO(2n))$. Then

$$H^*(BO(2n); Z) = Z[p_1, \ldots, p_n] + \text{2-torsion} \, .$$

Also

$$H^*(BO(2n+1); Z) = Z[p_1, \ldots, p_n] + \text{2-torsion} \, .$$

Denote also by $p_k \in H^{4k}(BSO(n); Z)$ the image of p_k under $H^*(BO(n); Z) \to H^*(BSO(n); Z)$. One new class has also to be defined, namely the Euler class. Under $\varrho: H^*(BU(2n); Z) \to H^*(BSO(2n); Z)$, let $W_{2n} = \varrho(c_n) \in H^{2n}(BSO(2n); Z)$. Then according to BOREL-HIRZE-BRUCH [7, p. 373], $H^*(BSO(2n+1); Z) = Z[p_1, \ldots, p_n] + 2\text{-torsion}$, $H^*(BSO(2n); Z) = Z[p_1, \ldots, p_{n-1}, W_{2n}] + 2\text{-torsion}$.

Suppose now that $\tau: E \to X$ is a bundle of n-dimensional vector spaces, with X a CW complex. There is a continuous inner product on the bundle [22, p. 37], so that we may consider τ an $O(n)$-bundle. There is then a bundle map

$$
\begin{array}{ccc}
E & \xrightarrow{f} & E' \\
\tau \downarrow & & \downarrow \\
X & \xrightarrow{\bar{f}} & BO(n)
\end{array}
$$

where $E' \to BO(n)$ is a universal $O(n)$-bundle with fiber R^n. The homomorphism

$$\bar{f}^*: H^*(BO(n); Z_2) \to H^*(X, Z_2)$$

is independent of the particular bundle map, by universality. The Whitney classes $w_k(\tau) \in H^k(X; Z_2)$ of the bundle τ are defined by $w_k(\tau) = \bar{f}^*(w_k)$. Similarly there are the Pontryagin classes of a vector space bundle, and Chern classes of complex vector space bundles.

Consider the embedding $O(m) \times O(n) \subset O(m+n)$ which identifies $\alpha \times \beta$ with the matrix $\begin{pmatrix} \alpha & 0 \\ 0 & \beta \end{pmatrix}$. It is not difficult to show that the induced $\varrho: H^*(BO(m+n); Z_2) \to H^*(BO(m) \times BO(n); Z_2)$ is given by $\varrho(w_k) = \Sigma_{p+q=k} w_p \otimes w_q$. When converted into a statement concerning vector space bundles, this becomes the classical Whitney sum theorem: for vector space bundles $\tau_i: E_i \to X$, $i = 1, 2$, we have $w_k(\tau_1 \otimes \tau_2) = \Sigma_{p+q=k} w_p(\tau_1) w_q(\tau_2)$. The Pontryagin classes reduced modulo an odd prime p also obey such a rule.

11. The Thom spaces

In this section we consider the Thom spaces of $SO(n)$-bundles, in preparation for the homotopy interpretation of the bordism groups.

The category of spaces with base point is useful here. The objects of the category are pairs (X, x_0) consisting of a space X and a base point $x_0 \in X$; abbreviate the pair simply by X. The maps of the category are the maps $(X, x_0) \to (Y, y_0)$. There is the sum $X \vee Y = X \times y_0 \cup x_0 \times Y \subset X \times Y$ with base point (x_0, y_0); $X \vee Y$ is a disjoint union of X and Y with base points identified.

If $X \supset A$ then X/A is the space obtained by identifying A to a point; the point corresponding to A is the base point. There is the product $X \wedge Y = X \times Y / X \vee Y$. In the category of spaces with base point, the circle S^1 is taken to be $I/\dot{I}$. The suspension SX of X is taken to be $X \wedge S^1$. Also $S^n = S(S^{n-1})$; it is true that S^n is homeomorphic to the n-dimensional sphere.

Shift now to the category of $SO(n)$ bundles (n variable) and bundle maps; the fiber is always the solid n-ball $C^n \subset R^n$. Denote a bundle by $\xi : E(\xi) \to B(\xi)$, and the union of the boundary spheres of the fibers of ξ by $\dot{E}(\xi)$. THOM has described a functor T from the category of $SO(n)$ bundles to the category of spaces with base point. Namely, T assigns to ξ the *Thom space* $T(\xi) = E(\xi)/\dot{E}(\xi)$, and T assigns to the bundle map $f : E(\xi) \to E(\eta)$ the map $T(f) : T(\xi) \to T(\eta)$ induced by $f : (E(\xi), \dot{E}(\xi)) \to \to (E(\eta), \dot{E}(\eta))$.

It may be seen that if $B(\xi)$ is a CW complex then $T(\xi)$ is also a CW complex, with no cells in dimension $< n$ except the base point.

Let $\xi \times \eta : E(\xi) \times E(\eta) \to B(\xi) \times B(\eta)$ be the product of ξ and η. Then $\xi \times \eta$ is a $SO(m) \times SO(n)$ bundle with fiber $C^m \times C^n$. Identify $C^m \times C^n$ with C^{m+n}, and consider $\xi \times \eta$ a $SO(m+n)$-bundle. Then

$$(E(\xi) \times E(\eta))^{\cdot} \cong \dot{E}(\xi) \times E(\eta) \cup E(\xi) \times \dot{E}(\eta)$$

$$T(\xi \times \eta) \cong T(\xi) \wedge T(\eta) .$$

In particular, if O_1 is the trivial line interval bundle over $B(\xi)$, then $T(\xi \oplus O_1) = T(\xi) \wedge S^1 = S T(\xi)$.

Let $\eta_k : E_k \to BSO(k)$ denote a universal $SO(k)$-bundle. The Thom space $T(\eta_k)$ is denoted by $MSO(k)$. Now $\eta_k \oplus O_1$ is an $SO(k+1)$-bundle over $BSO(k)$. There is then a bundle map $E(\eta_k \oplus O_1) \to E(\eta_{k+1})$ and an induced map $SMSO(k) \to MSO(k+1)$, unique up to homotopy type. More generally there is a map $MSO(p) \wedge MSO(q) \to MSO(p+q)$.

THOM introduced the spaces above in order to convert Ω_n into a stable homotopy group. We give now a bare outline of this noted theorem.

(11.1) **Thom.** *For* $k \geq n+2$ *we have* $\Omega_n \cong \pi_{n+k}(MSO(k))$, *and* $\mathfrak{N}_n \cong \pi_{n+k}(MO(k))$.

The proof is long [40, 22]. Here we merely define the isomorphisms. Suppose that M^n is a closed oriented n-manifold. Embed M^n in S^{n+k} via the Whitney embedding theorem. Denote by $\xi : A \to M^n$ the normal cell bundle to M^n in S^{n+k}. Assuming S^{n+k} oriented, we may assume the tangent bundle μ to S^{n+k} oriented; moreover the tangent bundle τ to M^n is oriented. Hence the normal bundle ξ is oriented so that the orientation of τ followed by the orientation to ξ yields the orientation of μ. We may then consider ξ an $SO(k)$-bundle. By § 10, A can be identified with a closed tubular neighborhood of M^n in S^{n+k}; identify the two. There is a bundle map

$$\begin{array}{ccc} A & \xrightarrow{f} & E(\eta_k) \\ \xi \downarrow & & \downarrow \eta_k \\ M^n & \xrightarrow{\bar{f}} & BSO(k) \end{array}$$

and an induced map $A/\dot{A} \to E/\dot{E}$ of Thom spaces. There is then the composite map f given by

$$S^{n+k} \to S^{n+k}/(S^{n+k}\backslash \operatorname{Int} A) \cong A/\dot{A} \xrightarrow{T(f)} E/\dot{E} = MSO(k) ,$$

and f represents an element of $\pi_{n+k}(MSO(k))$. Thom proved that the correspondence $[M^n] \to [f]$ is well-defined and an isomorphism. Here the transverse regularity concept was particularly essential.

We also point out the beginnings of the analysis of $H^*(MSO(n); Z)$. Consider an $SO(n)$-bundle $\xi: A \to X$ with fiber C^n and CW base space X. We have $H^i(A/\dot{A}; Z) \cong H^i(A, \dot{A}; Z)$, $i > 0$. An adaptation of the usual spectral sequence of a map yields a spectral sequence with $E_2^{p,q} = H^p(X; H^q(C^n, S^{n-1}; Z))$ and with E^∞ associated with a filtration of $H^{p+q}(A, \dot{A}; Z)$. Then $E_2^{p,q} = 0$ for $q \neq n$, $E_2^{p,n} = H^p(X; Z)$. We thus receive, since the spectral sequence has a single non-zero fiber degree, the Thom isomorphism [39],

$$(11.2) \qquad \Psi: H^p(X; Z) \cong H^{p+n}(A, \dot{A}; Z) .$$

Moreover Ψ is canonical. Given a bundle map

$$\begin{array}{ccc} A & \xrightarrow{f} & A' \\ \xi \downarrow & & \downarrow \xi' \\ X & \xrightarrow{\bar{f}} & X' \end{array}$$

there is a commutative diagram

$$\begin{array}{ccc} H^{p+n}(A, \dot{A}; Z) & \xleftarrow{f^*} & H^{p+n}(A', \dot{A}'; Z) \\ \Psi \uparrow & & \uparrow \Psi \\ H^p(X; Z) & \xleftarrow{\bar{f}^*} & H^p(X'; Z) . \end{array}$$

We have also the following theorem of Thom [40].

(11.3) *The map* $SMSO(k) \to MSO(k+1)$ *defined earlier in this section induces an isomorphism* $\pi_i(SMSO(k)) \cong \pi_i(MSO(k+1))$ *in dimensions* $i \leq 2k$.

Proof. As earlier denote the universal bundle by η_k and η_{k+1}. There is then the diagram of the bundle maps

$$\begin{array}{ccc} E(\eta_k \oplus O_1) & \xrightarrow{f} & E(\eta_{k+1}) \\ \downarrow & & \downarrow \\ BSO(k) & \xrightarrow{\bar{f}} & BSO(k+1) \end{array}$$

and the diagram

$$\begin{array}{ccc} H^{p+k+1}(SMSO(k); K) & \xleftarrow{T(f)^*} & H^{p+k+1}(MSO(k+1); K) \\ \cong \Big\uparrow \Psi & & \cong \Big\uparrow \Psi \\ H^p(BSO(k); K) & \xleftarrow{\bar{f}^*} & H^p(BSO(k+1); K) \end{array}$$

It follows from the resume of characteristic classes in § 10 that $\bar{f}^*(w_i)$ $= w_i$ for $i \leq k$ and coefficients Z_2. Hence with $K = Z_2$, $\bar{f}^*$ is an isomorphism in dimensions $\leq k$ and $T(f)^*$ is an isomorphism in dimensions $\leq 2k + 1$. By SERRE's map C theory [35], $\pi_i(SMSO(k)) \to$ $\to \pi_i(MSO(k+1))$ is an isomorphism modulo the class of odd torsion groups for $i \leq 2k$. A similar analysis for $K = Z_p$, p and odd prime, shows that $\pi_i(SMSO(k)) \to \pi_i(MSO(k+1))$ is an isomorphism modulo the class of torsion groups with no p-torsion for $i \leq 2k$. Hence $\pi_i(SMSO(k)) \to$ $\to \pi_i(MSO(k+1))$ is an isomorphism for $i \leq 2k$.

12. Homotopy interpretation of the bordism groups

THOM opened the way for a complete analysis of Ω_n by demonstrating an isomorphism $\Omega_n \cong \pi_{n+k}(MSO(k))$, $k \geq n + 2$. In this section we prepare for the study of the structure of $\Omega_n(X, A)$ by extending THOM's result to an isomorphism

$$\Omega_n(X, A) \cong \pi_{n+k}((X/A) \wedge MSO(k))$$

for (X, A) a CW pair. Always in this section pairs (X, A) will be CW pairs.

It is easiest to consider first the absolute case. Suppose that (M^n, f) is an oriented singular manifold in the space X. Embed M^n in S^{n+k}, $k \geq n + 2$. There is a tubular neighborhood N of M^n in S^{n+k} as in § 10, and N can be considered as the oriented normal k-cell bundle $\xi : N \to M^n$ to M^n. Let $BSO(k)$ denote a classifying space for $SO(k)$, chosen to be a countable CW complex, and let $\eta_k : E_k \to BSO(k)$ denote the universal oriented k-cell bundle. By a theorem of WHITEHEAD [44], $BSO(k)$ is then of the homotopy type of a locally finite CW complex. Hence we may take $BSO(k)$ to be locally finite. Then for any CW complex X, $X \times BSO(k)$ is a CW complex [44]. There is then a bundle map

$$\begin{array}{ccc} N & \xrightarrow{g} & E_k \\ \xi \downarrow & & \downarrow \eta_k \\ M^n & \xrightarrow{\bar{g}} & BSO(k) \, . \end{array}$$

There is also the map $f\xi : N \to M^n \to X$.
Hence we have

$$(f\xi) \times g : (N, \dot{N}) \to (X \times E_k, X \times \dot{E}_k) \, .$$

Denoting the map of quotient spaces by the same name, we have

$$(f\xi) \times g : N/\dot{N} \to X \times E_k/X \times \dot{E}_k \, .$$

We then have the composition

$$S^{n+k} \to S^{n+k}/(S^{n+k}\backslash \mathrm{Int}\, N) = N/\dot{N} \to X \times E_k/X \times \dot{E}_k$$

which we denote by h.

(12.1) *The homotopy class in $\pi_{n+k}(X \times E_k/X \times \dot{E}_k)$ of the map h above is a function only of the bordism class $[M^n, f]$ in $\Omega_n(X)$.*

Proof. Suppose (M_0, f_0) and (M_1, f_1) are oriented singular manifolds in X, and that each M_i^n is embedded in S^{n+k}. There are the tubular neighborhoods $\xi_i : N_i \to M_i^n$ in S^{n+k}, and bundle maps

$$\begin{array}{ccc} N_i & \xrightarrow{g_i} & E_k \\ \downarrow & & \downarrow \\ M_i^n & \xrightarrow{\bar{g}_i} & BSO(k). \end{array}$$

There is then

$$(f_i \xi_i) \times g_i : N_i/\dot{N}_i \to X \times E_k/X \times \dot{E}_k$$

and the induced map

$$h_i : S^{n+k} \to X \times E_k/X \times \dot{E}_k \,.$$

Suppose now that $(-M_0^n, f_0) \cup (M_1^n, f_1)$ bords. There is then a compact oriented manifold B^{n+1} with $\dot{B}^{n+1} = M_1^n \cup -M_0^n$ and a map $f : B^{n+1} \to X$ with $f|M_i^n = f_i$. Consider the oriented manifold $I \times S^{n+k}$, with M_0^n embedded in $0 \times S^{n+k} \cong S^{n+k}$ and M_1^n, embedded in $1 \times S^{n+k} \cong$ $\cong S^{n+k}$. As in section 10, B^{n+1} can be embedded in $I \times S^{n+k}$ with

$$B^{n+1} \cap (0 \times S^{n+k}) = M_0^n,\ B^{n+1} \cap (1 \times S^{n+k}) = M_1^n\,.$$

We may also suppose as in section 10 that there is a $0 < t_0 < 1$ such that if $(0, x) \in B^{n+1}$ then $(t, x) \in B^{n+1}$ for $0 < t < t_0$, and similarly for points $(1, x) \in B^{n+1}$. Suppose also that $I \times S^{n+k}$ is given the product metric.

For ε sufficiently small, there exists a tubular neighborhood N of radius ε of B^{n+1} in $I \times S^{n+k}$. We may suppose that N_0 and N_1 were also of radius ε. Then

$$N \cap (0 \times S^{n+k}) = N_0,\ N \cap (1 \times S^{n+k}) = N_1\,.$$

N can be identified with the oriented normal cell bundle $\xi : N \to B^{n+1}$ to B^{n+1} in $I \times S^{n+k}$. Moreover ξ restricted to M_i^n is ξ_i. Hence there exists a bundle map $g : N \to E_k$ with $g = g_i$ on N_i. There is $(f\xi) \times g : N \to$ $\to X \times E_k$ inducing

$$h : I \times S^{n+k} \to X \times E_k/X \times \dot{E}_k \,.$$

Moreover $h(0, x) = h(x)$ and $h(1, x) = h_1(x)$. The result then follows.

We thus receive a well-defined function

$$\tau : \Omega_n(X) \to \pi_{n+k}(X \times E_k/X \times \dot{E}_k), \quad k \geqq n + 2\,.$$

(12.2) τ *is a homomorphism.*

For consider oriented singular manifolds (M_0^n, f_0) and (M_1^n, f_1) in X. We have only to embed M_0^n in the interior of the lower hemisphere of S^{n+k}, and M_1^n in the interior of the upper hemisphere; we have then an embedding of the disjoint union $(M_1^n \cup M_2^n, f_1 \cup f_2)$ in S^{n+k}. One proceeds easily through the definition of τ to the conclusion.

Observe also that if X is a point p, then τ becomes the Thom isomorphism $\Omega_n \cong \pi_{n+k}(MSO(k))$ of (11.1). Thus we have the following.

(12.3) *For X a single point, τ is an isomorphism.*

Moreover if $\varphi : X \to Y$ is a map, there is the commutative diagram

$$
\begin{array}{ccc}
\Omega_n(X) & \xrightarrow{\ \tau\ } & \pi_{n+k}(X \times E_k / X \times \dot{E}_k) \\
\downarrow{\scriptstyle \varphi_*} & & \downarrow{\scriptstyle (\varphi_n \, id)_*} \\
\Omega_n(Y) & \xrightarrow[\ \tau\ ]{} & \pi_{n+k}(Y \times E_k / Y \times \dot{E}_k)\ .
\end{array}
$$

Recall the convention that $X/\emptyset$ is the disjoint union of X and a point, denoted here by ∞. Then

$$
(X/\emptyset) \wedge MSO(k) = (X \times E_k \cup \infty \times E_k)/X \times \dot{E}_k \cup \infty \times E_k) \cong
$$
$$
\cong X \times E_k / X \times \dot{E}_k\ .
$$

Using this identification, we can recast τ in the following form. Given (M^n, f) and the embedding of M^n in S^{n+k}, there is $f\xi : N \to X \subset X/\emptyset$ and $g : N/\dot{N} \to E_k/\dot{E}_k = MSO(k)$, giving a map

$$
(f\xi) \wedge g : N/\dot{N} \to (X/\emptyset) \wedge MSO(k)\ .
$$

Then $\tau[M, f]$ is represented by the composition

$$
S^{n+k} \to S^{n+k}/(S^{n+k} \backslash \mathrm{Int}\, N) = N/\dot{N} \xrightarrow{\ (f\xi) \wedge g\ } (X/\emptyset) \wedge MSO(k)\ .
$$

(12.4) *There exists a unique homomorphism*

$$
\tau : \Omega_n(X, x_0) \to \pi_{n+k}(X \wedge MSO(k)), \quad k \geq n + 2
$$

with commutativity holding in

$$
\begin{array}{ccc}
\Omega_n(X) & \longrightarrow & \Omega_n(X, x_0) \\
\downarrow{\scriptstyle \tau} & & \downarrow{\scriptstyle \tau} \\
\pi_{n+k}((X/\emptyset) \wedge MSO(k)) & \to & \pi_{n+k}(X \wedge MSO(k))\ .
\end{array}
$$

Proof. There is the sequence

$$
0 \to x_0 \cup \infty \to X \cup \infty \to X \cup \infty/x_0 \cup \infty = X \to 0
$$

inducing the exact sequence

$$
0 \to (x_0 \cup \infty) \wedge MSO(k) \to (X/\emptyset) \wedge MSO(k) \to X \wedge MSO(k) \to 0\ .
$$

Since $MSO(k)$ is a CW complex with no cells of dimension $< k$ except the base point, so then the three spaces of the above diagram are CW complexes with no cells of dimensions $< k$ except the base point. But there is the Blakers-Massey theorem [3]: If X and A are $(k-1)$-connected CW complexes, then the natural map $\pi_i(X, A) \to \pi_i(X/A)$ is an isomorphism for $i \leq 2k - 2$. It then follows from exactness of the ordinary homotopy sequence that $\pi_{n+k}((x_0/\emptyset) \wedge MSO(k)) \to$ $\to \pi_{n+k}((X/\emptyset) \wedge MSO(k)) \to \pi_{n+k}(X \wedge MSO(k))$ is exact. From the diagram

$$0 \to \Omega_n(x_0) \to \Omega_n(X) \to \Omega_n(X, x_0) \to 0$$

$$\pi_{n+k}((x_0/\emptyset) \wedge MSO(k)) \to \pi_{n+k}((X/\emptyset) \wedge MSO(k)) \to \pi_{n+k}(X \wedge MSO(k))$$

we obtain the conclusion.

Using the identification $\tilde{\Omega}_n(X) \cong \Omega_n(X, x_0)$, we thus obtain a homomorphism $\tau : \tilde{\Omega}_n(X) \to \pi_{n+k}(X \wedge MSO(k))$.

Recall that $S(X \wedge MSO(k)) \cong X \wedge SMO(k)$. Also denote by S the usual suspension homomorphism $\pi_i(X) \to \pi_{i+1}(SX)$; S assigns to the homotopy class of $f : S^i \to X$ the homotopy class of $f \wedge id : S^i \wedge S^1 = S^{i+1} \to X \wedge S^1 = SX$. We use the map $SMSO(k) \to MSO(k+1)$ of § 11 in the following remark.

(12.5) *Commutativity holds in*

$$\Omega_n(X)$$

$$\tau_1 \qquad\qquad \tau_2$$

$$\pi_{n+k}((X/\emptyset) \wedge MSO(k)) \to \pi_{n+k+1}((X/\emptyset) \wedge SMSO(k)) \to \pi_{n+k+1}((X/\emptyset) \wedge MSO(k+1))$$

Proof. Suppose that (M^n, f) is an oriented singular n-manifold in X. Embed M^n in S^{n+k}, let $\xi : N \to M^n$ denote a tubular neighborhood of M^n in S^{n+k}, and consider a bundle map

$$
\begin{array}{ccc}
N & \xrightarrow{g} & E_k \\
\downarrow & & \downarrow \\
M^n & \longrightarrow & BSO(k) \,.
\end{array}
$$

There is then

$$(f\xi) \times g : (N, \dot{N}) \to (X \times E_k, X \times \dot{E}_k) \subset ((X \cup \infty) \times E_k, X \times \dot{E}_k \cup \infty \times E_k).$$

The induced map

$$h : S^{n+k} \to (X/\emptyset) \wedge MSO(k)$$

represents $\tau_1[M^n, f]$.

Now S^{n+k} is embedded in S^{n+k+1} as the set of $(x_1, \ldots, x_{n+k+2})$ with $x_{n+k+2} = 0$. There is the tubular neighborhood $\xi' : N \to M^n$ of M^n in S^{n+k+1}, an oriented $(k+1)$-cell bundle. It is seen that ξ' is the Whitney sum $\xi \oplus O_1$, where O_1 is a trivial line bundle. There is then the diagram

$$N = E(\xi') = E(\xi \oplus O_1) \xrightarrow{g \oplus \mathrm{id}} E(\eta_k \oplus O_1) \longrightarrow E(\eta_{k+1}) = E_{k+1}$$

$$\begin{array}{ccc} \downarrow & \downarrow & \downarrow \\ M^n \xrightarrow{\quad\quad \bar{g} \quad\quad} BSO(k) \longrightarrow BSO(k+1) \end{array}$$

where $\eta_k : E_k \to BSO(k)$ is the universal bundle. Passing to Thom spaces, we get a diagram

$$N'/\dot{N'} \longrightarrow E(\eta_k \oplus O_1)/\dot{E}(\eta_k \oplus O_1) \to E(\eta_{k+1})/\dot{E}(\eta_{k+1})$$

$$\begin{array}{ccc} \downarrow{\cong} & \downarrow{\cong} & \downarrow{=} \\ (N/\dot{N}) \wedge S^1 \xrightarrow{g \wedge \mathrm{id}} MSO(k) \wedge S^1 \to MSO(k+1) \end{array}$$

The commutativity follows readily.

(12.6) *Using the identification*

$$S(X \wedge MSO(k)) = (X \wedge S^1) \wedge MSO(k) = SX \wedge MSO(k)$$

we get a commutative diagram

$$\tilde{\Omega}_n(X) \xrightarrow{\tau} \pi_{n+k}(X \wedge MSO(k))$$

$$\begin{array}{ccc} \downarrow & & \downarrow \\ \tilde{\Omega}_{n+1}(SX) \xrightarrow{\tau} \pi_{n+k+1}(SX \wedge MSO(k)). \end{array}$$

Proof. Suppose (M^n, f) represents an element of $\tilde{\Omega}_n(X)$. There is then a compact $(n+1)$-manifold B^{n+1} with $\dot{B}^{n+1} = M^n$. As in section 6, $S[M^n, f]$ is represented by (W^{n+1}, F) in $\tilde{\Omega}_{n+1}(SX)$, where $W^{n+1} = (B^{n+1} \times I)^{\cdot}$, where $F|M^n \times I = f \times id$, F maps $B^{n+1} \times 1$ and $B^{n+1} \times 0$ into the base point of SX.

Suppose M^n is embedded in S^{n+k}. There is $\xi : N \to M^n$, $g : N \to E_k$, $(f \xi) \wedge g : N \to X \wedge MSO(k)$ inducing $h : S^{n+k} \to X \wedge MSO(k)$. Then $\tau[M, f] = [h]$.

Using the Whitney embedding theorem as seen in section 10, we can embed W^{n+1} in S^{n+k+1} so that $M^n \times [1/2, 1] \cup B^{n+1} \times 1$ lies in the upper hemisphere, $M^n \times [0, 1/2] \cup B^{n+1} \times 0$ lies in the lower hemisphere, and $W^{n+1} \cap S^{n+k} = M^n \times 1/2$, which is identified with M^n. We may also suppose that W^{n+1} is orthogonal to S^{n+k} at their points of intersection. Let $\xi' : N' \to W^{n+1}$ be a tubular neighborhood of W^{n+1} in S^{n+k+1}, chosen so that $N' \cap S^{n+k} = N$ and so that ξ' restricted to $M^n \times 1/2 \cong M^n$ is ξ. We now have the maps

$$h : S^{n+k} \to N/\dot{N} \xrightarrow{(f\xi) \wedge g} X \wedge MSO(k) \,,$$

$$h' : S^{n+k} \to N'/\dot{N'} \xrightarrow{(F\xi') \wedge g'} SX \wedge MSO(k) \,.$$

We may consider $SX \wedge MSO(k)$ as a union of two cones, say C_+ and C_-, with $C_+ \cap C_- = X \wedge MSO(k) \subset SX \wedge MSO(k)$. The map h' is seen to map the upper hemisphere of S^{n+k+1} into C_+, the lower hemisphere in C_-, and h' agrees with h on S^{n+k}. It then follows that $[h'] = S[h]$, and the remark follows.

For every CW pair (X, A), consider the sequence

$$\ldots, (X/A) \wedge MSO(k), \quad (X/A) \wedge MSO(k+1), \ldots$$

together with the maps

$$((X/A) \wedge MSO(k)) \to (X/A) \wedge SMSO(k) \to (X/A) \wedge MSO(k+1) .$$

(12.7) *The composition homomorphism* $\pi_{n+k}((X/A) \wedge MSO(k)) \overset{S}{\longrightarrow}$ $\overset{S}{\longrightarrow} \pi_{n+k+1}((X/A) \wedge SMSO(k)) \to \pi_{n+k+1}((X/A) \wedge MSO(k+1))$ *is an isomorphism for* $k \geq n + 2$.

Proof. Recall that $MSO(k)$ has no cells of dimension $< k$ except for the base point. The same then follows for $(X/A) \wedge MSO(k)$. Hence $(X/A) \wedge MSO(k)$ is $(k-1)$-connected. Then S is an isomorphism for $k \geq n + 2$ by the stability theorems.

Consider now the map $SMSO(k) \to MSO(k+1)$ of section 10. If η_l denotes the universal bundle, there is a bundle map

$$
\begin{array}{ccc}
E = E(\eta_k \oplus O_1) & \overset{F}{\longrightarrow} & E(\eta_{k+1}) = E' \\
\downarrow & & \downarrow \\
BSO(k) & \overset{\bar{F}}{\longrightarrow} & BSO(k+1) .
\end{array}
$$

There is the Thom diagram (11.2)

$$
\begin{array}{ccc}
H^n(BSO(k+1)) & \overset{\bar{F}^*}{\longrightarrow} & H^n(BSO(k)) \\
\cong \downarrow \Psi & & \cong \downarrow \Psi \\
H^{n+k+1}(E', \dot{E}') & \overset{F^*}{\longrightarrow} & H^{n+k+1}(E, \dot{E}) .
\end{array}
$$

For coefficients Z_2, $\bar{F}^*$ maps Whitney class w_i into Whitney class w_i, $1 \leq i \leq k$, while $\bar{F}^*$ kills w_{k+1}. Hence $\bar{F}^*$ is an isomorphism for $n \leq k$, as is F^*. For coefficients Z_k, p odd, $\bar{F}^*$ maps Pontryagin class into Pontryagin class. However for k even the Euler class in $H^k(BSO(k))$ may not be in the image. However $\bar{F}^*$ is an isomorphism for $n < k$ as then is F^*. Hence for every prime p

$$H^{n+k+1}(E', \dot{E}') \to H^{n+k+1}(E, \dot{E})$$

is an isomorphism over Z_p for $n < k$, as is

$$H^{n+k+1}(MSO(k+1)) \to H^{n+k+1}(SMSO(k)) .$$

Now $\tilde{H}^*(P \wedge Q; K) \cong \tilde{H}^*(P; K) \otimes \tilde{H}^*(Q; K)$, for K a field. It then follows that

$$H^{n+k+1}((X/A) \wedge MSO(k+1)) \to H^{n+k+1}((X/A) \wedge SMSO(k))$$

is an isomorphism, coefficients Z_p, for $n < k$. Hence

$$\pi_{n+k+1}((X/A) \wedge S\,MSO(k)) \to \pi_{n+k+1}((X/A) \wedge MSO(k+1))$$

is an isomorphism modulo the class of torsion groups having every element of order prime to p, $n \leq k-2$, and this for every p. Hence it is an isomorphism and the assertion follows.

Now for a CW pair (X, A), define $\mathscr{H}_n(X, A) = \pi_{n+k}((X/A) \wedge MSO(k))$, $k \geq n+2$, where the groups are identified for different values of k by the isomorphism of (11.7).

There is a boundary operator $\partial : \mathscr{H}_n(X, A) \to \mathscr{H}_{n-1}(A, B)$ for triples $X \supset A \supset B$ defined by means of the Blakers-Massey theorem. Namely the natural map

$$\pi_{n+k}((X/B) \wedge MSO(k),\, (A/B) \wedge MSO(k)) \to \pi_{n+k}((X/A) \wedge MSO(k))$$

is an isomorphism for $k \geq n+2$, since the spaces involved are $(k-1)$-connected. The boundary operator

$$\partial : \pi_{n+k+1}((X/B \wedge MSO(k),\, (A/B) \wedge MSO(k)) \to \pi_{n+k}((A/B) \wedge MSO(k))$$

then becomes the boundary

$$\partial : \mathscr{H}_{n+1}(X, A) \to \mathscr{H}_n(A, B) \, .$$

The reader may verify that the resulting ∂ is well-defined.

Maps $\quad \varphi : (X, A) \to (X_1, A_1) \quad$ clearly induce homomorphisms $\varphi_* : \mathscr{H}_n(X, A) \to \mathscr{H}_n(X_1, A_1)$.

(12.8) *The covariant functor $\{\mathscr{H}_n(X, A),\, \varphi_*,\, \partial\}$ satisfies the first six Eilenberg-Steenrod axioms.*

The general study of such homology theories has been carried out by G. W. WHITEHEAD [43]; our procedure here has been suggested by his results. There are, as in section 6, such consequences as reduced groups $\widetilde{\mathscr{H}}_n(X)$, etc. Here $\widetilde{\mathscr{H}}_n(X) \cong \mathscr{H}_n(X, x_0) \cong \pi_{n+k}(X \wedge MSO(k))$. There is also a suspension isomorphism $S : \widetilde{\mathscr{H}}_n(X) \to \widetilde{\mathscr{H}}_{n+1}(SX)$, the ordinary suspension $\pi_{n+k}(X \wedge MSO(k)) \to \pi_{n+k+1}(SX \wedge MSO(k))$. The composition

$$\mathscr{H}_{n+1}(X, A) \cong \widetilde{\mathscr{H}}_{n+1}(X/A) \xrightarrow{\varphi_*} \widetilde{\mathscr{H}}_{n+1}(SA) \xrightarrow{S^{-1}} \widetilde{\mathscr{H}}_n(A) \to \mathscr{H}_n(A)$$

is just $(-1)^{n-1}\partial$, as with $\Omega_*(X, A)$ (see (6.5) and (6.6)).

We have defined for every CW pair a homomorphism

$$\Omega_n(X, A) \cong \widetilde{\Omega}_n(X/A) \xrightarrow{\tau} \mathscr{H}_n(X/A) \cong \mathscr{H}_n(X, A)$$

following (12.4). The role of (12.5) is to prove that τ is well-defined; that is, independent of k. The role of (12.6) is to prove, in view of (6.5), (6.6) and the immediately preceding remarks, that commutativity holds in

$$\begin{array}{ccc} \Omega_n(X, A) & \xrightarrow{\ \tau\ } & \mathscr{H}_n(X, A) \\ \downarrow{\scriptstyle \partial} & & \downarrow{\scriptstyle \partial} \\ \Omega_{n-1}(A) & \xrightarrow{\ \tau\ } & \mathscr{H}_{n-1}(A) \, . \end{array}$$

If $\varphi : (X, A) \to (X_1, A_1)$, it is obvious that commutativity holds in

$$\begin{array}{ccc} \Omega_n(X, A) & \xrightarrow{\ \tau\ } & \mathscr{H}_n(X, A) \\ \downarrow{\scriptstyle \varphi_*} & & \downarrow{\scriptstyle \varphi_*} \\ \Omega_n(X_1, A_1) & \xrightarrow{\ \tau\ } & \mathscr{H}_n(X_1, A_1) \, . \end{array}$$

(12.9) Theorem. *The homomorphism τ of the homology theory $\{\Omega_*(X, A), \varphi_*, \partial\}$ into the homology theory $\{\mathscr{H}_*(X, A), \varphi_*, \partial\}$, where $\mathscr{H}_n(X, A) \cong \pi_{n+k}((X/A) \wedge MSO(k))$. $k \geqq n + 2$, is an isomorphism of the two theories over the category of CW pairs (X, A).*

Proof. From the Thom result (12.3), τ is an isomorphism for a point, that is, on the coefficient groups. It will follow that τ is an isomorphism for all finite CW pairs (X, A). For suppose τ is an isomorphism on all pairs (X, A) where X has $\leqq k$ cells. Suppose now that X has $k + 1$ cells. If A also has $k + 1$ cells then $X = A$ and $\Omega_n(X, A) = \mathscr{H}_n(X, A) = 0$. If A has $\leqq k$ cells, select a subcomplex X_1 with $X_1 \supset A$ and with X_1 having k cells. There is

$$\begin{array}{ccccccc} \cdots \to & \Omega_n(X_1, A) & \to & \Omega_n(X, A) & \to & \Omega_n(X, X_1) & \to \cdots \\ & \downarrow{\scriptstyle \tau_1} & & \downarrow{\scriptstyle \tau_2} & & \downarrow{\scriptstyle \tau_3} & \\ & \mathscr{H}_n(X_1, A) & \to & \mathscr{H}_n(X, A) & \to & \mathscr{H}_n(X, X_1) & . \end{array}$$

Moreover τ_1 is an isomorphism by induction. Also $\Omega_n(X, X_1) \cong {} \cong \Omega_n(I^n, S^{n-1})$ and $\mathscr{H}_n(X, X_1) \cong \mathscr{H}_n(I^n, S^{n-1})$. An easy argument using the fact that τ is an isomorphism of coefficient groups shows τ_3 to be an isomorphism. The five lemma then shows τ_2 to be an isomorphism.

Hence τ is an isomorphism on finite CW pairs. For CW pairs (X, A), it is easy to see that

$$\Omega_n(X, A) \cong \operatorname{Dir\,Lim} \Omega_n(X_\alpha, A_\alpha) \, ,$$

over all finite pairs (X_α, A_α) in (X, A). Also

$$\pi_{n+k}(X, A) \cong \operatorname{Dir\,Lim} \pi_{n+k}(X_\alpha, A_\alpha) \, ,$$

and

$$\pi_{n+k}((X/A) \wedge MSO(k)) \cong \operatorname{Dir\,Lim} \pi_{n+k}((X_\alpha/A_\alpha) \wedge MSO(k)) \, .$$

That is,

$$\mathscr{H}_n(X, A) \cong \operatorname{Dir\,Lim} \mathscr{H}_n(X_\alpha, A_\alpha)$$

we thus see that τ is an isomorphism for every CW pair (X, A).

We sketch now WHITEHEAD's more general fashion for obtaining generalized homology theories. Here a *spectrum* M will be a sequence $\ldots, M_n, M_{n+1}, \ldots$ of countable CW complexes with base points, with M_n defined for all n sufficiently large, together with maps $SM_n \to M_{n+1}$, such that

(a) M_n is $(n-1)$-connected;

(b) $SM_n \to M_{n+1}$ induces isomorphisms $\pi_i(SM_n) \to \pi_i(M_{n+1})$ for $i < 2n$. We can then define, for (X, A) a CW pair, a group

$$\mathscr{H}_n(X,A) = \text{Dir Lim}\, \pi_{n+k}((X/A) \wedge M_k).$$

For X finite dimensional, we have

$$\mathscr{H}_n(X, A) = \pi_{n+k}((X/A) \wedge M), \quad k \text{ large.}$$

There is then a boundary operator $\partial : \mathscr{H}_n(X, A) \to \mathscr{H}_{n-1}(A, B)$, defined as above. There is thus defined the functor $\{\mathscr{H}_n(X, A), \varphi_*, \partial\}$, due to G. W. WHITEHEAD [43], satisfying the first six Eilenberg-Steenrod axioms.

There are cohomology groups $H^n(M; K) = \text{Dir Lim}\, H^{n+k}(M_k; K)$, and homotopy groups $\pi_n(M) = \text{Dir Lim}\, \pi_{n+k}(M_k)$. The coefficient groups $\mathscr{H}_n(p)$ of the homology theory are the groups $\pi_n(M)$.

In this section we have used the Thom spectrum $MSO : \ldots, MSO(n)$, $MSO(n+1), \ldots$ with the maps $SMSO(n) \to MSO(n+1)$.

There is another spectrum of particular importance to us. Given an abelian group π, there is a spectrum

$$K(\pi) : \cdots, K(\pi, n), K(\pi, n+1), \ldots$$

with maps $SK(\pi, n) \to K(\pi, n+1)$ as follows. There are the fundamental cohomology classes $\tau_m \in H^m(\pi, m; \pi)$, and the suspension homomorphism $S : H^i(\pi, n; \pi) \to H^{i+1}(SK(\pi, n); \pi)$. There exists a map $f : SK(\pi,n) \to K(\pi, n+1)$ with $f^*(\tau_{n+1}) = S(\tau_n)$, and any two such maps are homotopic. It is easy to verify conditions (a) and (b) of the definition of a spectrum. The Serre-Cartan theory [12] shows that $H^n(K(\pi); \pi')$ is isomorphic to the stable cohomology operations

$$H^k(X; \pi) \to H^{k+n}(X; \pi').$$

Consider now the homology theory based on the spectrum $K(\pi)$. The coefficient group $\mathscr{H}_n(p)$ is $\pi_n(K(\pi)) = \pi_{k+n}(K(\pi, k))$. Hence $\mathscr{H}_n(p) = 0$ for $n \neq 0$, $\mathscr{H}_0(p) \cong \pi$. By the Eilenberg-Steenrod uniqueness theorem for ordinary homology, we then have $\mathscr{H}_n(X, A) \cong H_n(X, A; \pi)$ for finite CW pairs (X, A).

13. Duality and cobordism

Following ATIYAH [1], we sketch a generalized cohomology theory $\{\Omega^n(X, A), \varphi^*, \delta\}$ defined via mappings into Thom spaces; the $\Omega^n(X, A)$

are the cobordism groups of (X, A). There is given the Thom-Atiyah duality theorem, of the type of Poincaré duality. We go on to show duality of the Alexander type. The section is not of fundamental importance to the remainder of this work, but is sketched for completeness.

We work in the category of CW complexes with base point. Denote by $[X, Y]$ the set of homotopy classes of maps $X \to Y$. The spaces $MSO(1)$, $MSO(2), \ldots, MSO(k), \ldots$ together with the maps $SMSO(k) \to \to MSO(k + 1)$ define the Thom spectrum MSO. Following ATIYAH, the nth cobordism group of (X, A) is defined by

$$\Omega^n(X, A) = \mathrm{Dir\ Lim}\,[S^k(X/A), MSO(k + n)]$$

for every integer n, where (X, A) is a finite CW pair. The homomorphisms of the direct system are the composites $[X, MSO(n)] \xrightarrow{S} \xrightarrow{S} [SX, SMSO(n)] \to [SX, MSO(n + 1)]$.

(13.1) *The homomorphisms* $[X, MSO(n)] \to [SX, MSO(n + 1)]$ *are isomorphisms whenever X is a CW complex of dimension $\leq 2n - 2$.*

Proof. That $S: [X, MSO(n)] \to [SX, SMSO(n)]$ is an isomorphism if $\dim X \leq 2n - 2$ follows from the well-known stability theorems [3], since $MSO(n)$ is $(n - 1)$-connected. That $[SX, SMSO(n)] \to \to [SX, MSO(n + 1)]$ is an isomorphism follows from (11.3) and a remark of SPANIER [36, Appentix].

It follows from (13.1) that

$$\Omega^n(X, A) = [S^k(X/A), MSO(k + n)]$$

for k sufficiently large. The group $\Omega^n(X)$ is defined by $\Omega^n(X) = \Omega^n(X, \emptyset)$. For any map $f: (X, A) \to (Y, B)$ there is an induced homomorphism $f^*: \Omega^n(Y, B) \to \Omega^n(X, A)$. In the fashion of stable homotopy there is a coboundary $\delta: \Omega^n(A) \to \Omega^{n+1}(X, A)$. For further details the reader may consult ATIYAH [1].

(13.2) **Atiyah.** *The cobordism functor $\{\Omega^n(X, A), \varphi^*, \delta\}$ satisfies the first six Eilenberg-Steenrod axioms for a cohomology theory. For a single point, $\Omega^n(p) = \Omega_{-n}$.*

There are reduced groups $\tilde{\Omega}^n(X)$. These may be defined as $\tilde{\Omega}^n(X) = \Omega^n(X)/Im\,\varepsilon^*$ where $\varepsilon: X \to p$ collapses X to a point. We leave it as an exercise to show a canonical isomorphism

$$\tilde{\Omega}^n(X) \cong [S^k X, MSO(k + n)], \quad k \text{ large}.$$

For a pair (X, A) there is an exact sequence

$$\cdots \to \Omega^k(X, A) \to \tilde{\Omega}^k(X) \to \tilde{\Omega}^k(A) \to \Omega^{k+1}(X, A) \to \cdots$$

and the commutative diagram

$$\Omega^k(X, A) \to \tilde{\Omega}^k(X) \to \tilde{\Omega}^k(A)$$
$$\downarrow s \qquad\qquad \downarrow s$$
$$\Omega^{k+1}(SX, SA) \to \tilde{\Omega}^{k+1}(SX) \to \tilde{\Omega}^{k+1}(SA) .$$

We turn now to duality.

(13.3) **Thom-Atiyah.** *If (X, A) is a finite CW pair with $X\backslash A$ an oriented n-manifold without boundary, then there is a canonical isomorphism $u : \Omega_k(X\backslash A) \cong \Omega^{n-k}(X, A)$.*

We outline the proof of (13.3). Suppose first $k < n/2$. By the embedding theorems, then $[V^k, f] \in \Omega_k(X - A)$ may be represented as the inclusion of a closed oriented regular submanifold $V^k \subset X\backslash A$. A tubular neighborhood N of V^k in $X\backslash A$ can be identified with the normal $(n - k)$-cell bundle to V^k, $\xi : N \to V^k$. The fibers of N are oriented so that the orientation of the tangent bundle of V^k followed by the orientation of ξ yields the orientation of the tangent bundle to $X\backslash A$ restricted to V. There is a diagram of bundle maps

$$N \xrightarrow{f} E$$
$$\xi \downarrow \qquad \downarrow \eta$$
$$V^k \xrightarrow{\bar{f}} BSO(n - k) .$$

There is then the map

$$X/A \to N/\dot{N} \to E/\dot{E} = MSO(n - k)$$

which represents an element of $[X/A, MSO(n - k)] = \Omega^{n-k}(X, A)$. This defines u in the stable range. In the stable range commutativity holds in

$$\Omega_k(X\backslash A) \xrightarrow{u} \Omega^{n-k}(X, A)$$
$$\downarrow i_* \qquad\qquad \downarrow s$$
$$\Omega_k(SX\backslash SA) \xrightarrow{u} \Omega^{n-k+1}(SX, SA)$$

where i_* is induced by inclusion.

In arbitrary dimensions the isomorphism u is defined as the composition

$$\Omega_k(X\backslash A) \underset{\cong}{\xrightarrow{i_*}} \Omega_k(S^rX\backslash S^rA) \underset{\cong}{\xrightarrow{u}} \Omega^{n+r-k}(S^rX, S^rA) \underset{\cong}{\longleftarrow} \Omega^{n-k}(X, A)$$

for r large.

Consider now a finite simplicial complex X embedded as a subcomplex of S^n. For such complexes there is the duality theory of SPANIER-WHITEHEAD [37]. In particular, any finite complex $D_nX \subset S^n\backslash X$ which is a deformation retract of $S^n\backslash X$ is an *n-dual* of X. If X is a finite CW complex, there is a finite simplicial complex X' of the same homotopy type as X [44]. An n-dual D_nX' is defined to be a *weak n-dual D_nX* of X.

(13.4) *If X is a finite CW complex with a weak n-dual $D_n X$, there is a canonical isomorphism*

$$u' : \tilde{\Omega}_k(X) \cong \tilde{\Omega}^{n-k-1}(D_n X) .$$

Proof. We may confine ourselves to finite simplicial complexes X embedded as proper subcomplexes of S^n. Since X is contractible to a point in S there is a short exact sequence $0 \to \tilde{\Omega}^{n-k-1}(X) \to$ $\to \Omega^{n-k}(S^n, X) \to \tilde{\Omega}^{n-k}(S^n) \to 0$. There is also the exact sequence

$$0 \to \tilde{\Omega}_k(S^n\backslash X) \to \Omega_k(S^n\backslash X) \to \Omega_k \to 0 .$$

Duality yields a diagram

$$0 \to \tilde{\Omega}_k(S^n\backslash X) \to \Omega_k(S^n\backslash X) \to \Omega_k \to 0$$
$$\downarrow \qquad\qquad\qquad \downarrow$$
$$0 \to \tilde{\Omega}^{n-k-1}(X) \to \Omega^{n-k}(S^n, X) \to \tilde{\Omega}^{n-k}(S^n) \to 0 .$$

There is then a unique isomorphism $\tilde{\Omega}_k(S^n\backslash X) \cong \tilde{\Omega}^{n-k-1}(S^n, X)$ such that commutativity holds. Since $\tilde{\Omega}_k(D_n X) = \tilde{\Omega}_k(S^n\backslash X)$, we get an isomorphism $u' : \tilde{\Omega}_k(D_n X) \cong \tilde{\Omega}^{n-k-1}(X)$, which is sufficient to show (13.4).

CHAPTER II

Computation of the bordism groups

In the previous chapter, we have defined and characterized geometrically the homology theory $\{\Omega_*(X, A), \varphi_*, \partial\}$ of bordism. Thus the stage is set for their computation, at least in many cases. In order to compute, we use the powerful results of MILNOR (Ω_* has no odd torsion) and WALL (see section 14) on $MSO(k)$. In section 14 we prove that the bordism spectral sequence is trivial modulo the class of odd torsion groups. In section 15 it is proved that if X has no odd torsion then $\Omega_n(X) = \Sigma_{p+q=n} H_p(X; \Omega_q)$; in section 18 it is shown that if X has no torsion then $\Omega_*(X) \cong H_*(X; Z) \otimes \Omega$ as an Ω-module.

Generalizing the Stiefel-Whitney numbers and the Pontryagin numbers of a manifold, in section 17 we define natural numerical invariants of maps $f : M^n \to X$. These are functions only of the bordism class of f. If all torsion of X consists of elements of order two, the bordism class of f is determined by the Whitney numbers and the Pontryagin numbers of f.

14. Triviality mod C

Denote by C the class of torsion groups having all elements of odd order. The fundamental result of this chapter is the following.

(14.1) **Theorem.** *For any CW pair (X, A) the bordism spectral sequence is trivial* $\mathrm{mod}\, C$.

The purpose of this section is to prove (14.1). We must show that the image of each $d^r : E^r_{p,q} \to E^r_{p-r,q+r-1}$ is an odd torsion group. We use in a basic way the following theorem of C. T. C. WALL [42], which is now assumed.

(14.2) **Wall.** *The module $H^*(MSO(k); Z_2)$ over the* mod 2 *Steenrod algebra is isomorphic in dimensions* $< 2k$ *to a direct sum of Steenrod algebras $H^*(Z, m_i; Z_2)$ and $H^*(Z_2, n_i; Z_2)$.*

Put in terms of spectra of § 12, $H^*(MSO; Z_2)$ is isomorphic as a module over the Steenrod algebra to a direct sum of copies of $H^*(K(Z); Z_2)$ and of $H^*(K(Z_2); Z_2)$.

Proof of (14.1). Let $a \in H^{m_i}(MSO; Z_2)$ denote the generator of one of the submodules of $H^*(MSO; Z_2)$ isomorphic to $H^*(K(Z); Z_2)$. The Bockstein $S_q^1 : H^{m_i}(MSO; Z_2) \to H^{m_i+1}(MSO; Z_2)$ kills a. Hence the integral Bockstein $H^{m_i}(MSO; Z_2) \to H^{m_i+1}(MSO; Z)$ maps a into an element a', a' of order two, which is zero when restricted mod 2. Hence $a' = 2b$ for some $b \in H^{m_i+1}(MSO; Z)$. But additively $H^*(MSO(k); Z) \cong$ $\cong H^*(BSO(k); Z)$ by the Thom isomorphism (11.2), and hence all 2-torsion consists of elements of order two by the results in § 10. Since $2a' = 0$, $a' = 2b$ it follows that $a' = 0$. Thus a is the restriction of an integral class $\alpha \in H^{m_i}(MSO; Z)$.

The elements α, a are represented by unique elements $\alpha_k \in$ $\in H^{k+m_i}(MSO(k); Z)$ and $a_k \in H^{k+m_i}(MSO(k); Z_2)$ for $k > m_i$. For each $k > m_i$ there is a cellular map $f_k : MSO(k) \to K(Z, m_i + k)$, unique up to homotopy, with $f^*(\tau) = \alpha_k$, $\tau \in H^{m_i+k}(Z, m_i + k; Z)$ the fundamental class. Also $f^*(\tau \bmod 2) = a_k$.

The diagram

$$
\begin{array}{ccc}
SMSO(k) & \longrightarrow & MSO(k+1) \\
\downarrow{\scriptstyle Sf_k} & & \downarrow{\scriptstyle f_{k+1}} \\
SK(Z, m_i + k) & \longrightarrow & K(Z, m_i + k + 1) \, ,
\end{array}
$$

where the horizontal maps are the spectrum maps, is then seen to be commutative up to homotopy.

Consider now a variant of the homology theory of § 12 based on the spectrum $K(Z)$. Define

$$
{}^i\! K_s(X, A) = \mathrm{Dir}\,\mathrm{Lim}\,\pi_{s+k}((X/A) \wedge K(Z, m_i + k)) \, .
$$

It follows from § 12 that ${}^i\! K_s(X, A) = H_{s-m_i}(X, A; Z)$. There is a spectral sequence $\{{}^i\! E^r_{p,q}\}$ for the homology theory ${}^i\! K_s(X, A)$. We have ${}^i\! E^1_{p,q} = {}^i\! K_{p+q}(X^p/X^{p-1}) \cong H_{p+q-m_i}(X^p, X^{p-1}; Z)$. Hence ${}^i\! E^1_{p,q} = 0$ if $q \neq m_i$, ${}^i\! E^2_{p,m_i} = C_p(X, A)$. It is also the case that ${}^i\! E^2_{p,m_i} = H_p(X, A; Z)$. Since there is just one non-zero fiber degree, the spectral sequence is trivial for $r \geq 2$. The maps $f_k : MSO(k) \to K(Z, m_i + k)$, all k, induce canonical homomorphisms $f_{k*} : \Omega_k(X, A) \to {}^iK_k(X, A)$, and homomorphisms of the bordism spectral sequence $\{E_{p,q}^r\}$ into the spectral sequence $\{{}^iE_{p,q}^r\}$.

In WALL's decomposition of $H^*(MSO; Z_2)$, consider next a submodule isomorphic to $H^*(K(Z_2); Z_2)$. Let $b \in H^{n_j}(MSO; Z_2)$ be a generator and $b_k \in H^{k+n_j}(MSO(k); Z_2)$ a representative. There are maps $g_k : MSO(k) \to K(Z_2, n_j + k)$ with $g_k^*(\tau) = b_k$, τ the fundamental class.

Define a homology theory ${}^jK(X, A)$ by

$$ {}^jK_s(X, A) = \mathrm{Dir\ Lim}\, \pi_{s+k}((X/A) \wedge K(Z_2, n_j + k)) . $$

It follows from § 12 that ${}^jK_s(X, A) \cong H_{s-n_j}(X, A; Z_2)$. There is a spectral sequence $\{{}^jE_{p,q}^r\}$ for the homology theory ${}^jK(X, A)$ with ${}^jE_{p,q}^1 = H_{p+q-n_j}(X^p, X^{p-1}; Z_2)$. Hence ${}^iE_{p,q}^1 = 0$ if $q \neq n_j$, and ${}^jE_{p,n_j}^2 = H_p(X, A; Z_2)$. The spectral sequence is trivial for $r \geq 2$. There are canonical homomorphisms $g_* : \Omega_k(X, A) \to {}^jK_k(X, A)$, and homomorphisms

$$ \{E_{p,q}^r\} \to \{{}^jE_{p,q}^r\} . $$

Define now a homology theory $K_*(X, A)$ by

$$ K_k(X, A) = \Sigma_i\, {}^iK_k(X, A) + \Sigma_j\, {}^jK_k(X, A) . $$

We have a canonical homomorphism $\theta : \Omega_k(X, A) \to K_k(X, A)$, the direct sum of the f_* and the g_*. There is also a spectral sequence $\{'E_{p,q}^r\}$ with $'E_{p,q}^r = \Sigma_i\, {}^iE_{p,q}^r + \Sigma_j\, {}^jE_{p,q}^r$, and with d^r the direct sum of the ${}^id^r$ and the ${}^jd^r$. Then $'E_{p,q}^1 = K_{p+q}(X^p, X^{p-1})$. There is moreover a homomorphism of $\{E_{p,q}^r\}$ into $\{'E_{p,q}^r\}$ with $E_{p,q}^1 \to 'E_{p,q}^1$ the homomorphism $\theta : \Omega_{p+q}(X^p, X^{p-1}) \to K_{p+q}(X^p, X^{p-1})$.

We must now analyze the homomorphism θ. For p a point, $\Omega_n(p) = \pi_{n+k}(MSO(k))$ for k large. Fix k large, and let $P_i = K(Z, m_i + k)$, $Q_i = K(Z, n_i + k)$. Then

$$ K_n(p) = \Sigma_i \pi_{n+k}(K(Z, m_i + k)) + $$
$$ + \Sigma_j \pi_{n+k}(K(Z_2, n_j + k)) \cong \pi_{n+k}(\Pi P_i \times \Pi Q_j) . $$

The homomorphism $\theta : \Omega_n(p) \to K_n(p)$ is induced by the map $f : MSO(k) \to \Pi P_i \times \Pi Q_j$ defined by $f = \Pi f_k^i \times \Pi g_k^j$. According to the theorem of WALL, $f^* : H^*(\Pi P_i \times \Pi Q_j; Z_2) \to H^*(MSO(k); Z_2)$ is an isomorphism in dimensions $< 2k$. Hence $f_* : \pi_{n+k}(MSO(k)) \to \pi_{n+k}(\Pi P_i \times \Pi Q_j)$ is an isomorphism modulo the class C of odd torsion groups if $n \leq k - 2$ [35]. Hence $\theta : \Omega_n(p) \to K_n(p)$ is an isomorphism mod C. It follows now from the proof of (12.9) that θ is an isomorphism mod C for any CW pair (X, A).

Coming back now to the homomorphism $\{E_{p,q}^r\} \to \{'E_{p,q}^r\}$ we recall that $\{'E_{p,q}^r\}$ is trivial for $r \geq 2$. Moreover $\theta : E_{p,q}^1 \to 'E_{p,q}^1$ is an isomorphism mod C. Hence $E^r_{p,q} \to {}'E^r_{p,q}$ is an isomorphism mod C for $r \geqq 1$, and $\{E^r_{p,q}\}$ is trivial mod C for $r \geqq 2$.

(14.2) **Theorem.** *For any CW pair (X, A) there is an isomorphism* $\theta : \Omega_n(X, A) \cong \Sigma_{p+q=n} H_p(X, A; \Omega_q) \bmod C$. *For any finite CW complex X there is an isomorphism* $\Omega^m(X) \cong \Sigma_{p-q=m} H^p(X; \Omega_q) \bmod C$.

Proof. In the proof of (14.1) we have constructed a mod C isomorphism $\theta : \Omega_n(X, A) \cong K_n(X, A) \bmod C$ where $K_n(X, A) = \Sigma H_{n-m_i}(X, A; Z) + \Sigma H_{n-n_j}(X, A; Z_2)$. Now $K_n(X, A) = \Sigma_{p+q=n} H_p(X, A; \Lambda_q)$ where Λ_q is the direct sum of as many copies of Z as there are i with $m_i = q$ and as many Z_2 as there are j with $n_j = q$. In particular $K_n(p) = \Lambda_n$ and there is the mod C isomorphism $\theta : \Omega_n(p) \to K_n(p)$, or $\theta : \Omega_n \to \Lambda_n$. But Ω_n and Λ_n are finitely generated abelian groups all of whose torsion is 2-torsion; here we assume the theorem of MILNOR that Ω_n has no odd torsion [26]. If two such groups are mod C isomorphic, they are isomorphic. Hence $\Lambda_q \cong \Omega_q$ and

$$\Omega_n(X, A) \cong \Sigma_{p+q=n} H_p(X, A; \Omega_q) \bmod C .$$

The result for $\Omega^m(X)$ follows from duality arguments, and we leave it to the reader.

15. Steenrod representation

We return to the homomorphism $\mu : \Omega_n(X, A) \to H_n(X, A; Z)$ of section 6. To each oriented singular manifold $[B^n, f]$ in (X, A), μ assigns the image of the orientation class of B^n under the homomorphism $f_* : H_n(B^n, \dot{B}^n; Z) \to H_n(X, A; Z)$. The image of μ is the subgroup of integral homology classes representable in the sense of STEENROD. We can now make progress in the study of μ, using the fact (7.2) that the image of μ is the set of permanent cycles of the term $E^2_{n,0}$ of the bordism spectral sequence.

(15.1) **Theorem.** *If (X, A) is a CW pair then the bordism spectral sequence collapses (is trivial) if and only if $\mu : \Omega_n(X, A) \to H_n(X, A; Z)$ is an epimorphism for all $n \geqq 0$.*

Proof. The spectral sequence collapses if $d^r : E^r_{p,q} \to E^r_{p-r,q+r-1}$ is trivial for all $r \geqq 2$. It is clear that if the spectral sequence collapses, then $\mu : \Omega_n(X, A) \to H_n(X, A; Z)$ is an epimorphism for all $n \geqq 0$.

Assume now that μ is an epimorphism. Then $d^r : E^r_{n,0} \to E^r_{n-r,r-1}$ is trivial for all r. Consider the operation of Ω on the bordism spectral sequence as in section 7. We have $H_p(X, A; Z) \otimes \Omega_q = E^2_{p,0} \otimes \Omega_q \to E^2_{p,q} = H_p(X, A; \Omega_q)$. From (7.1) this is a monomorphism with image $H_p(X, A; Z) \otimes \Omega_q \subset H_p(X, A; \Omega_q)$. Since every element of $E^2_{p,0}$ is a permanent cycle, so is every element of $H_p(X, A; Z) \otimes \Omega_q \subset H_p(X, A; \Omega_q)$. Now $H_p(X, A; Z) \otimes \Omega_q$ is a direct summand of $E^2_{p,q}$, the other summand $T_{p,q}$ being isomorphic to $\mathrm{Tor}(H_{p-1}(X, A; Z), \Omega_q)$. Since Ω_q has no odd torsion, $T_{p,q}$ consists of 2-torsion only. By (14.1), d^2 carries $E^2_{p,q}$ onto an odd torsion group, and since $T_{p,q}$ consists solely of 2-torsion, then $d^2(T_{p,q}) = 0$. Thus $d^2 = 0$. As we continue through the spectral sequence, it is seen to be trivial. The theorem then follows.

Thom has shown that in $H_7(Z_3 + Z_3, 1; Z)$ there is an integral class which is not Steenrod representable [40]. The bordism spectral sequence of $K(Z_3 + Z_3, 1)$ is thus non-trivial.

(15.2) **Theorem.** *If (X, A) is a CW pair such that each $H_n(X, A; Z)$ is finitely generated and has no odd torsion, then the bordism spectral sequence is trivial. Moreover $\mu: \Omega_n(X, A) \to H_n(X, A; Z)$ is an epimorphism and $\Omega_n(X, A) \cong \Sigma_{p+q=n} H_p(X, A; \Omega_q)$.*

Proof. The bordism spectral sequence is trivial mod C and $E^2_{p,q} = H_p(X, A; \Omega_q)$ has no odd torsion, thus $d^2: E^2_{p,q} \to E^2_{p-2,q+1}$ is trivial. Hence $E^3_{p,q} \cong E^2_{p,q}$, and we repeat the argument, eventually showing the spectral sequence trivial. Since no $E^2_{p,q}$ has odd torsion, then $\Omega_*(X, A)$ has no odd torsion.

Now from (14.2) we have a mod C isomorphism $\Omega_n(X, A) \cong \cong \Sigma_{p+q=n} H(X, A; \Omega_q)$ joining finitely generated groups without odd torsion. The torsion subgroup of $\Omega_n(X, A)$ must then map isomorphically onto the torsion subgroup of $\Sigma H_p(X, A; \Omega_q)$. The groups also have the same rank, hence $\Omega_n(X, A) \cong \Sigma_{p+q=n} H_p(X, A; \Omega_q)$.

(15.3) **Theorem.** *Let (X, A) be a CW pair. For each homology class $c \in H_n(X, A; Z)$ there is a integer k with $(2k + 1) c$ Steenrod representable.*

Proof. It follows from (14.1) that the image R of $\mu: \Omega_n(X, A) \to E^2_{n,0} = H_n(X, A; Z)$ has $H_n(X, A; Z)/R$ an odd torsion group. Hence (15.3) follows; this result improves the result of Thom that some multiple of c is representable. The following is obtained in the same fashion.

(15.4) *Let (X, A) be a CW pair. Every element of the 2-torsion subgroup of $H_n(X, A; Z)$ is Steenrod representable.*

There are also gap theorems for the bordism spectral sequence, which we examine next.

(15.5) *For a CW pair (X, A), $E^r_{p,q}$ consists entirely of elements of order 2 if $q \not= 0 \bmod 4$.*

Proof. As was noted in § 2, the results of Milnor and Wall show that Ω_q consists entirely of elements of order 2 if $q \not= 0 \bmod 4$. Now $E^2_{p,q} = H_p(X, A; Z) \otimes \Omega_q + \mathrm{Tor}(H_{p-1}(X, A; Z), \Omega_q)$ and the remark follows for $E^2_{p,q}$, and hence $E^r_{p,q}$ in view of (14.1).

(15.6) *The boundary $d^r: E^r_{p,q} \to E^r_{p-r,q+r-1}$ is trivial unless $q = 0 \bmod 4$ and $r = 1 \bmod 4$.*

This follows from (14.1) and (15.5). We come now to the gap theorem.

(15.7) **Theorem.** *Let (X, A) be a CW pair. If for each pair $r > s$ with $r - s = 1 \bmod 4$, either $H_r(X, A; Z)$ is a 2-torsion group or $H_s(X, A; Z)$*

has no odd torsion, then every element of $H_n(X, A; Z)$ is Steenrod representable.

Proof. By (15.6), $E^5_{p,q} \cong E^2_{p,q}$, and $E^2_{p,q} = H_p(X, A; Z) \otimes \Omega_q +$ $+ \mathrm{Tor}(H_{p-1}(X, A; Z), \Omega_q)$. If $H_p(X, A; Z)$ is a 2-torsion group, so is $E^5_{p,q}$ and by (14.1) $d^5 : E^5_{p,q} \to E^5_{p-5,q+4}$ is trivial. If $H_{p-5}(X, A; Z)$ has no odd torsion, then neither does $E^5_{p-5,q+4}$, so $d^5 : E^5_{p,q} \to E^5_{p-5,q+4}$ is again trivial. Thus $E^2_{p,q} \cong E^5_{p,q} \cong E^6_{p,q} \cong E^9_{p,q}$ and we repeat the argument inductively, showing that the spectral sequence collapses.

16. A generalization of Rochlin's theorem

There is the homomorphism $r : \Omega_n \to \mathfrak{N}_n$ obtained by ignoring the orientation of a closed oriented manifold; that is, $r[M^n] = [M^n]_2$. It was shown by Rochlin that the sequence $\Omega_n \xrightarrow{2} \Omega_n \xrightarrow{r} \mathfrak{N}_n$ is exact [34]. We show in this section that for any CW pair the corresponding sequence

$$\Omega_n(X, A) \xrightarrow{2} \Omega_n(X, A) \to \mathfrak{N}_n(X, A)$$

is also exact, where $r[B^n, f] = [B^n, f]_2$.

We first make some remarks concerning unoriented bordism. Let $\eta : E' \to BO(k)$ denote a universal k-cell bundle with structural group $O(k)$. There is a Thom space $MO(k) = E'/\dot{E}'$. Just as in section 12, there is an isomorphism

$$\mathfrak{N}_n(X, A) \cong \pi_{n+k}((X/A) \wedge MO(k))$$

for k large. There is also a diagram

$$
\begin{array}{ccc}
E & \xrightarrow{\;v\;} & E' \\
\eta \downarrow & & \downarrow \eta \\
BSO(k) & \xrightarrow{\;\bar{v}\;} & BO(k)
\end{array}
$$

of $O(k)$-bundle maps, since $E \to BSO(k)$ can be considered an $O(k)$-bundle. We get an induced map $v : E/\dot{E} \to E'/\dot{E}'$, that is, a map $MSO(k) \to MO(k)$. From the properties of characteristic classes, it is seen that $\bar{v}^* : H^*(BO(k); Z_2) \to H^*(BSO(k); Z_2)$ is an epimorphism. By the Thom isomorphism (11.2), $v^* : H^*(MO(k); Z_2) \to H^*(MSO(k); Z_2)$ is also an epimorphism. We leave the following as an exercise.

(16.1) *The diagram*

$$
\begin{array}{ccc}
\Omega_n(X, A) & \to & \pi_{n+k}((X/A) \wedge MSO(k)) \\
r \downarrow & & \downarrow (\mathrm{Id} \wedge v)_* \\
\mathfrak{N}_n(X, A) & \to & \pi_{n+k}((X/A) \wedge MO(k))
\end{array}
$$

is commutative for k large.

We proceed now to the proof of the theorem.

(16.2) **Theorem.** *For a CW pair (X, A) the sequence $\Omega_n(X, A) \xrightarrow{2}$ $\xrightarrow{2} \Omega_n(X, A) \xrightarrow{r} \mathfrak{N}_n(X, A)$ is exact.*

Proof. Since $\mathfrak{N}_n(X, A)$ has every element of order two, $Im\, 2 \subset Ker$. Suppose now that $x \in \Omega_n(X, A)$ has $r(x) = 0$. The order of proof is to show that the mod C isomorphism $\theta : \Omega_n(X, A) \to K_n(X, A)$ of the proof of (14.1) has $\theta(x)$ divisible by two. Since θ is a mod C isomorphism, x will then be divisible by two. In order to show $\theta(x)$ divisible by two, it will be sufficient to show that the image of x under each of the homomorphisms $\Omega_n(X, A) \to {}^iK_n(X, A)$, $\Omega_n(X, A) \to {}^jK_n(X, A)$ is divisible by two. We proceed to show this for the homomorphism $\Omega_n(X, A) \to$ $\to {}^iK_n(X, A)$.

Recall that the definition of ${}^iK_n(X, A)$ was as follows, for k large. One started with a certain $a_k \in H^{m_i + k}(MSO(k); Z_2)$, the restriction of an integral class $\alpha_k \in H^{m_i + k}(MSO(k); Z)$. There was defined a map $f_k : MSO(k) \to K(Z, m_i + k)$ with $f_k^*(\tau) = \alpha_k$, so that $f^*(\tau \bmod 2) = a_k$. Then ${}^iK_n(X, A) = \pi_{n+k}((X/A) \wedge K(Z, m_i + k)) \cong H_{n-m_i}(X, A; Z)$.

Since $\bar{v}^* : H^*(MO(k); Z_2) \to H^*(MSO(k); Z_2)$ is an epimorphism, there is a $c_k \in H^{m_i + k}(MO(k); Z_2)$ with $\bar{v}^*(c_k) = a_k$. Define $h_k : MO(k) \to$ $\to K(Z_2, m_i + k)$ so that $h_k^*(\tau') = c_k$ where τ' is the fundamental class. There is the group ${}^iL_n(X, A) = \pi_{n+k}((X/A) \wedge K(Z_2, m_i + k)) \cong$ $\cong H_{n-m_i}(X, A; Z_2)$ and the homomorphism $\mathfrak{N}_n(X, A) \to {}^iL_n(X, A)$ induced by h_k. There is also a map $w : K(Z, m_i + k) \to K(Z_2, m_i + k)$ with $w^*(\tau') = \tau \bmod 2$. It is seen that the diagram

$$
\begin{array}{ccc}
MSO(k) & \xrightarrow{\; f_k \;} & K(Z, m_i + k) \\
\downarrow{\scriptstyle v} & & \downarrow{\scriptstyle w} \\
MO(k) & \xrightarrow{\; h_k \;} & K(Z_2, m_i + k)
\end{array}
$$

is homotopy commutative. Hence

$$
\begin{array}{ccc}
\pi_{n+k}((X/A) \wedge MSO(k)) & \xrightarrow{(id \wedge f_k)_*} & \pi_{n+k}((X/A) \wedge K(Z, m_i + k)) \\
\downarrow{\scriptstyle (id \wedge v)_*} & & \downarrow{\scriptstyle (id \wedge w)_*} \\
\pi_{n+k}((X/A) \wedge MO(k)) & \xrightarrow{(id \wedge h_k)_*} & \pi_{n+k}((X/A) \wedge K(Z_2, m_i + k))
\end{array}
$$

is commutative. We next see that

$$
\begin{array}{ccc}
\Omega_n(X, A) & \xrightarrow{\; \theta_i \;} & {}^iK_n(X, A) \cong H_{n-m_i}(X, A; Z) \\
\downarrow{\scriptstyle r} & \downarrow & \downarrow{\scriptstyle s} \\
\mathfrak{N}_n(X, A) & \longrightarrow & {}^iL_n(X, A) \cong H_{n-m_i}(X, A; Z_2)
\end{array}
$$

is commutative, where s is restriction mod 2. We see that $(id \wedge w)_*$ yields a canonical homomorphism of the homology theory ${}^iK_*$ into

$^iL_*$, and that for (X, A) the pair $(p, \emptyset)$, it is for $n = m_i$ the natural homomorphism $Z \to Z_2$. But $^iK_*$ and $^iL_*$ are homology theories with a single non-zero coefficient group. The Eilenberg-Steenrod uniqueness theorem [19, p. 100] then applies to show that if two homomorphisms coincide on the coefficient group, they coincide in general. Hence the diagram is commutative.

Recall now that $x \in \Omega_n(X, A)$ was supposed to have $r(x) = 0$. Hence $\theta_i(x) \in {}^iK_n(X, A) \cong H_{n-m_i}(X, A; Z)$ has $s\theta_i(x) = 0$. By exactness of $H_{n-m_i}(X, A; Z) \xrightarrow{2} H_{n-m_i}(X, A; Z) \xrightarrow{s} H_{n-m_i}(X, A; Z_2)$, we have $\theta_i(x)$ divisible by two. We have then that in $\theta : \Omega_n(X, A) \to K_n(X, A)$, $\theta(x)$ is divisible by two. Since θ is a mod C isomorphism, it is not difficult to show x divisible by two.

17. Algebraic invariants of maps

In this section we generalize the well-known Stiefel-Whithney numbers and Pontryagin numbers of a manifold M^n, obtaining Whitney numbers and Pontryagin numbers of a map $f : M^n \to X$. We prove theorems concerning when these determine the bordism class $[M^n, f]$ in $\Omega_n(X)$.

Let M^n be a closed oriented manifold with orientation class $\sigma_n \in H_n(M^n; Z)$. Denote by $1 = w_0, w_1, \ldots, w_n$ the Stiefel-Whitney classes of M^n, that is the Whitney classes of the tangent bundle to M^n. To every partition $i_1 + \cdots + i_k = n$ there is the element $w_{i_1} \ldots w_{i_k} \in$ $\in H^n(M^n; Z_2)$ and the corresponding Stiefel-Whitney number $\langle w_{i_1} \ldots w_{i_k}, \sigma_n \rangle \in Z_2$. Here $\langle , \rangle$ denotes the dual Kronecker pairing of cohomology and homology.

If $[M^n]_2 = 0$, then all Stiefel-Whitney numbers of M^n vanish; this is due to PONTRYAGIN [33]. The proof is as follows. Let B^{n+1} be a compact manifold with $\dot{B}^{n+1} = M^n$. Let $\tilde{w}_0, \ldots, \tilde{w}_{n+1}$ denote the Stiefel-Whitney classes of B^{n+1}. The differentiable collaring theorem (1.2) shows that the bundle on M^n induced from the tangent bundle to B^{n+1} is the Whitney sum of the tangent bundle to M^n with a trivial line bundle. From the Whitney sum theorem and the naturality of Whitney classes, $i^*(\tilde{w}_i) = w_i$ where i is the inclusion $M^n \subset B^{n+1}$. Then $\langle w_{i_1} \ldots w_{i_k}, \sigma_n \rangle$ $= \langle i^* \tilde{w}_{i_1} \ldots i^* \tilde{w}_{i_k}, \sigma_n \rangle = \langle w_{i_1} \ldots w_{i_k}, i_* \sigma_n \rangle = 0$ since σ_n lies in the kernel of i_*. Thus the Stiefel-Whitney numbers of M^n depend only on $[M^n]_2$. THOM went on to show the following [40].

(17.1) **Thom.** *If M^n is a closed manifold, then $[M^n]_2 = 0$ if and only if all the Stiefel-Whitney numbers of M^n vanish.*

Next consider a map $f : M^n \to X$ where M^n is a closed manifold. We seek to associate numbers with $[M^n, f]_2 \in \mathfrak{N}_n(X)$ which are analogous to Stiefel-Whitney numbers. Let $h^m \in H^m(X; Z_2)$ be a cohomology class of X. For every partition $i_1 + \cdots + i_k = n - m$, the number $\langle w_{i_1} \ldots w_{i_k} f^*(h^m), \sigma_n \rangle \in Z_2$ is defined; we call this number a *Whitney* *number of the map f associated with h^m.* If h^0 is the unit class of $H^*(X;Z_2)$, then the Whitney numbers associated with h^0 are precisely the Stiefel-Whitney numbers of M^n.

Note that if $[M^n, f]_2 = 0$, then all the Whitney numbers of $f: M^n \to X$ vanish. Suppose B^{n+1} is a compact manifold with $\dot{B}^{n+1} = M^n$ and that $F: B^{n+1} \to X$ is a map with $F|M^n = f$. Then $\langle w_{i_1} \ldots w_{i_k} f^*(h^m), \sigma^n \rangle = \langle i^* \tilde{w}_{i_1} \ldots i^* \tilde{w}_{i_k} i^* F^*(h^m), \sigma_n \rangle = \langle \tilde{w}_{i_1} \ldots w_{i_k} F^*(h^m), i_* \sigma_n \rangle = 0$. Thus the Whitney classes of $f: M^n \to X$ only depend on $[M^n, f]_2 \in \mathfrak{N}_n(X)$.

Let X be a CW complex and for each n let $\{c_{n,i}\}$ be an additive base for $H_n(X;Z_2)$. According to (8.1), for each $c_{n,i}$ we may select a singular manifold $f_i: M_i^n \to X$ with $f_{i*}(\sigma_n) = c_{n,i}$. Define $h: H_*(X;Z_2) \otimes \otimes \mathfrak{N} \to \mathfrak{N}_*(X)$, an $\mathfrak{N}$-module homomorphism, by $h(c_{n,i} \otimes 1) = [M_i^n, f_i]_2 \in \mathfrak{N}_n(X)$. Since $H^*(X;Z_2) \otimes \mathfrak{N}$ is a free graded $\mathfrak{N}$-module, h is well-defined.

(17.1) **Theorem.** *The $\mathfrak{N}$-module homomorphism $h: H^*(X;Z_2) \otimes \otimes \mathfrak{N} \to \mathfrak{N}_*(X)$ is an isomorphism for each finite CW complex.*

Proof. Consider first the Whitney numbers of a product $[M_i^n, f_i]_2 [V^m]_2 = [M_i^n \times V^m, f_i \pi]$ where π is projection $M_i^n \times V^m \to M_i^n$. Let $c^{n,i} \in H^n(X;Z_2)$ be the cohomology class dual to $c_{n,i}$; that is, $\langle c^{n,i}, c_{n,i} \rangle = 1$ while $\langle c^{n,i}, c_{n,j} \rangle = 0$ for $i \neq j$. Select a partition $i_1 + \cdots + i_k = m$ and consider the Whitney numbers of $f_i \pi$ associated with $c^{n,i}$. Note that $(f_i \pi)^* c^{n,i} = f_i^* c^{n,i} \otimes 1$. Let $v_j \in H^j(V^m;Z_2)$ denote the Stiefel-Whitney classes of V^m. Any Stiefel-Whitney class of the product $M_i^n \times V^m$ is of form $W_j = 1 \otimes v_j +$ terms involving Stiefel-Whitney classes of M_i^n. From dimensional considerations we see that $W_{i_1} \ldots W_{i_k}(f_i \pi)^* c^{n,i} = f_i^* c^{n,i} \otimes v_{i_1} \ldots v_{i_k}$. Thus

$$\langle W_{i_1} \ldots W_{i_k}(f_i \pi)^* c^{n,i}, \sigma_n \times \sigma_m \rangle = \langle f^*(c^{n,i}), \sigma_n \rangle \langle v_{i_1} \ldots v_{i_k}, \sigma_m \rangle$$
$$= \langle v_{i_1} \ldots v_{i_k}, \sigma_m \rangle ,$$

a Stiefel-Whitney number of $[V^m]_2$. Furthermore if we use $c^{n,j}$ with $j \neq i$, the Whitney numbers of the product associated with $c^{n,j}$ all vanish.

Now suppose there is an expression $\Sigma_{m,i}[M_i^{n-m}, f_i]_2 [V_i^m]_2 = 0$. All the Whitney numbers must vanish. We show inductively that $[V_i^m]_2 = 0$ for all m, i. Suppose that for $m < m_0$ and all i it has been shown that $[V_i^m]_2 = 0$. We choose a partition $i_1 + \cdots + i_k = m_0$ and a c^{n-m_0, i_0}; then $\Sigma_{m \geq m_0, i} \langle W_{i_1} \ldots W_{i_k}(f_i \pi)^* c^{n-m_0, i_0}, M_i^{n-m} \times V_i^m \rangle = 0$.

Now note that it follows from dimensional considerations that $\langle W_{i_1} \ldots W_{i_k}(f_i \pi)^* c^{n-m_0, i_0}, M_i^{n-m} \times V_i^m \rangle = 0$ for all $m > m_0$ and all i. The point is that the expression for computing the Whitney number involves $f^*(c^{n-m_0, i_0}) \in H^{n-m_0}(M^{n-m};Z_2)$ which is 0 if $m > m_0$. Thus

$$\Sigma_i \langle W_{i_1} \ldots W_{i_k}(f_i \pi)^* c^{n-m_0, i_0}, M_i^{n-m_0} \times V_i^{m_0} \rangle = 0 .$$

However if $i \neq i_0$ the term is zero, while for $i = i_0$, we get the Stiefel-Whitney number $\langle v_{i_1} \ldots v_{i_k}, V_{i_0}^{m_0} \rangle$ of $V_{i_0}^{m_0}$. Hence the Stiefel-Whitney numbers of $V_{i_0}^{m_0}$ are all zero, and $[V_{i_0}^{m_0}]_2 = 0$. The induction is established, and hence h is a monomorphism.

For finite CW complexes X, (8.2) shows that the unoriented bordism spectral sequence is trivial, so that additively $H_*(X; Z_2) \otimes \mathfrak{N} \cong \mathfrak{N}_*(X)$. We then see by dimensional considerations that h is an isomorphism for finite CW complexes X. We have now fully established (8.3). In passing we have also shown the following.

(17.2) **Theorem.** *If $f : M^n \to X$ is an unoriented singular manifold in a finite CW complex X then $[M^n, f]_2 = 0$ if and only if all the Whitney numbers of $[M^n, f]_2$ vanish.*

We wish now to indicate a more functorial treatment of Whitney numbers. For a CW complex X denote by $\mathfrak{N}^n(X)$ the vector space of linear functionals from $\mathfrak{N}_n(X)$ to Z_2; that is, $\mathfrak{N}^n(X) = \mathrm{Hom}(\mathfrak{N}_n(X), Z_2)$. Consider the product $X \times BO(n)$ of X with the classifying space $BO(n)$. We define now a well defined homomorphism

$$H^n(X \times BO(n); Z_2) \to \mathfrak{N}^n(X) \,,$$

an epimorphism for all $n \geq 0$.

An element of $\mathfrak{N}_n(X)$ is determined by a map $f : M^n \to X$ where M^n is a closed manifold. The tangent bundle of M^n may be induced by a map $g : M^n \to BO(n)$. The product map $F = f \times g : M^n \to X \times BO(n)$ induces
$$F^* : H^n(X \times BO(n); Z_2) \to H^n(M^n; Z_2) \,.$$

For each $c^n \in H^n(X \times BO(n); Z_2)$ we consider $\langle F^*(c^n), \sigma_n \rangle \in Z_2$, σ_n the orientation class of M^n. It is seen that $\langle F^*(c^n), \sigma_n \rangle$ depends only on $[M^n, f]_2$ and on c^n. If c^n is fixed and $[M^n, f]_2$ varies over $\mathfrak{N}_n(X)$ we obtain an element of $\mathfrak{N}^n(X) = \mathrm{Hom}(\mathfrak{N}_n(X), Z_2)$. Thus we obtain a homomorphism $H^n(X \times BO(n); Z_2) \to \mathfrak{N}^n(X)$. In view of (17.2) it is an epimorphism.

Now this homomorphism has a non-trivial kernel $W^n(X) \subset$ $\subset H^n(X \times BO(n); Z_2)$. This kernel represents generalized Wu relations. That is, if X is a single point p, then the kernel of $H^n(\{p\} \times BO(n); Z_2) \to$ $\to \mathfrak{N}^n(p)$ is precisely the set of Wu relations as shown by DOLD [17].

For each $k \leq n$ and each partition $i_1 \leq \cdots \leq i_l$ of k containing no term of the form $2^j - 1$ (i.e. to each non-dyadic partition) there is associated an element $s(i_1, \ldots, i_l) \in H^k(BO(n); Z_2)$, a polynomial in the Whitney classes $w_1, \ldots, w_k$. Namely s is given by the symmetric function $\Sigma t_1^{i_1} \ldots t^{i_l}$. Let $S^k \subset H^k(BO(n); Z_2)$ be the subspace spanned by the $s(i_1, \ldots, i_l)$. Now additively $S^k \cong \mathfrak{N}_k \cong \mathfrak{N}^k(p)$ and in fact under $H^n(\{p\} \times BO(n); Z_2) \to \mathfrak{N}^n(p)$, S^n is carried isomorphically onto $\mathfrak{N}^n(p)$ [40].

In general let $S^n(X) \subset H^n(X \times BO(n); Z_2)$ be given by $S^n(X) = \Sigma_0^n H^{n-k}(X; Z_2) \otimes S^k$. The point is that under $H^n(X \times BO(n); Z_2) \to \mathfrak{N}^n(X)$ the subspace $S^n(X)$ is carried isomorphically onto $\mathfrak{N}^n(X)$. It is only necessary to show $S^n(X) \cap W^n(X) = 0$, but this is entirely analogous to (17.2). Thus we have a splitting $H^n(X \times BO(n); Z_2) = S^n(X) \oplus W^n(X)$. As we shall see, this may be regarded as a canonical sum decomposition.

Let $\varphi: X \to Y$ be a map; then $\varphi_*: \mathfrak{N}_n(X) \to \mathfrak{N}_n(Y)$ is defined and by duality $\Phi: \mathfrak{N}^n(Y) \to \mathfrak{N}^n(X)$ is also defined. Let $\varphi \times id: X \times BO(n) \to Y \times BO(n)$ induce $\Phi^*: H^n(Y \times BO(n); Z_2) \to H^n(X \times BO(n); Z_2)$. It is seen that the diagram

$$
\begin{array}{ccc}
H^n(X \times BO(n); Z_2) & \to & \mathfrak{N}^n(X) \\
\uparrow{\scriptstyle \Phi^*} & & \uparrow{\scriptstyle \Phi} \\
H^n(Y \times BO(n); Z_2) & \to & \mathfrak{N}^n(Y)
\end{array}
$$

commutes. By duality the kernel of Φ consists of those homomorphisms which annihilate the image of $\varphi_*: \mathfrak{N}_*(X) \to \mathfrak{N}_*(Y)$. The fact is that under $H^n(Y \times BO(n); Z_2) \to \mathfrak{N}^n(Y)$ the kernel of Φ^* is carried onto the kernel of Φ. Suppose $c_n \in H^n(Y \times BO(n); Z_2)$ goes into the kernel of Φ. We write $c_n = a_n + b_n$ where $a_n \in S^n(Y)$ and $b_n \in W^n(Y)$. Now $\Phi^*(c_n) \in W^n(X)$ but $\Phi^*(a_n) \in S^n(X)$ and $\Phi^*(b_n) \in W^n(X)$; hence $\Phi^*(a_n) = 0$, but a_n and c_n cover the same homomorphism in $\mathfrak{N}^n(Y)$. We get immediately as a corollary.

(17.3) *Let $\varphi: X \to Y$ be a map joining finite CW complexes. The necessary and sufficient condition $[M^n, f]_2 \in \mathfrak{N}_n(Y)$ lie in the image of $\varphi_*: \mathfrak{N}_n(X) \to \mathfrak{N}_n(Y)$ is that every Whitney number of $[M^n, f]_2$ associated with an element in the kernel of $\varphi^*: H^*(Y; Z_2) \to H^*(X; Z_2)$ must vanish.*

For the remainder of this section, we proceed by analogy to the oriented bordism groups $\Omega_n(X)$.

Let $w = (i_1, \ldots, i_k)$ be a partition, and $p_\omega = p_{i_1} \cdots p_{i_k}$ the cup-product of Pontryagin classes of the tangent bundle of a closed oriented manifold M^n. Let $h^m \in H^m(X; Z)$ be a cohomology class with $m = n - 4(i_1 + \cdots + i_k)$. For an oriented singular manifold $f: M^n \to X$ we then have the numbers $\langle p_\omega f^*(h^m), \sigma(M^n) \rangle \in Z$; we call these the *Pontryagin numbers* of the map f.

Since $\Omega_n(X) \cong \Sigma_{p+q=n} H_p(X; \Omega_q) \bmod C$ we see that $\Omega_n(X) \otimes Q \cong \Sigma_{p+q=n} H_p(X; \Omega_q \otimes Q)$ where Q denotes the rationals. This generalizes the theorem of Thom [40]. The following can be shown just as (17.2).

(17.4) **Theorem.** *Two oriented singular n-manifolds in X represent the same element in $\Omega_n(X) \otimes Q$ if and only if their corresponding Pontryagin numbers are equal.*

With this we can show that in many cases Whitney numbers and Pontryagin numbers determine the bordism class.

(17.5) Theorem. *Let X be a finite CW complex such that the torsion of $H_*(X;Z)$ consists of elements of order two. Then two oriented singular n-manifolds in X represent the same element in $\Omega_n(X)$ if and only if they have the same Whitney numbers and the same Pontryagin numbers.*

Proof. In view of (15.2), $\Omega_n(X) \cong \Sigma_{p+q=n} H_p(X;\Omega_q)$ and thus every torsion class in $\Omega_n(X)$ also has order two. Suppose all the Whitney numbers and the Pontryagin numbers of $[M^n, f]$ vanish. Then by (17.2), $[M^n, f]_2 = 0$. Thus by the exact sequence $\Omega_n(X) \to \Omega_n(X) \to \mathfrak{N}_n(X)$ of (16.2), $[M^n, f] = 2[V^n, g]$. However by (17.3) the class $[M^n, f]$ is a torsion class, as is $[V^n, g]$. Hence $2[V^n, g] = 0 = [M^n, f]$.

18. The existence of an Ω-base

We have seen in (8.3) and (17.2) that $\mathfrak{N}_*(X)$ always has a homogeneous $\mathfrak{N}$-base. We go on to show that in some cases there is a corresponding result for $\Omega_*(X)$.

(18.1) Theorem. *If X is a finite CW complex for which $H_*(X;Z)$ has no torsion, then $\Omega_*(X)$ has a homogeneous Ω-base and is thus a free graded Ω-module.*

Proof. Let $\{c_{n,i}\}$ denote an additive base for $H_*(X;Z)$. According to (15.2) there are oriented singular n-manifolds $f: M_i^n \to X$ with $f_*(\sigma) = c_{n,i}$. We shall show that $\{[M_i^n, f]\}$ forms a homogeneous Ω-base in $\Omega_*(X)$.

We rely on the triviality of the bordism spectral sequence. There is the filtration $\Omega_n(X) = J_{n,0} \supset \cdots \supset J_{0,n} \supset 0$ with $J_{k,n-k}/J_{k-1,n-k+1} \cong \cong E_{k,n-k}^{\infty} \cong E_{k,n-k}^2 = H_k(X;\Omega_{n-k})$.

Since X has no torsion, $H_k(X;Z) \otimes \Omega_{n-k} \cong H_k(X;\Omega_{n-k})$ and thus $E_{k,0}^2 \otimes \Omega_{n-k} \cong E_{k,n-k}^2$.

Let $A \subset \Omega_*(X)$ be the submodule generated by the $\{[M_i^n, f]\}$. We show by induction on k that $J_{k,n-k} \subset A$. Suppose this is true for $k-1$. Choose $\alpha \in J_{k,n-k}$. There is $\hat{\alpha} \in E_{k,n-k}^2$ corresponding to α. Since $E_{k,0}^2 \otimes \Omega_{n-k} \cong E_{k,n-k}^2$, $\hat{\alpha}$ is uniquely expressed as $\Sigma_i c_{k,i} \otimes [V_i^{n-k}]$. There is $\beta = \Sigma_i [M_i^k, f][V_i^{n-k}]$. We may take $f: M_i^k \to X$ as a map into the k-skeleton so that $\beta \in J_{k,n-k}$. Now β also corresponds to $\hat{\alpha} \in E_{k,n-k}^2$ thus $\alpha - \beta \in J_{k-1,n-k+1}$. By induction we see that $A = \Omega_*(X)$.

The independence can be seen as follows. If

$$\Sigma_i [M_i^{n-m}, f][V_i^m] = 0,$$

then we can see from (17.3) that all the $[V_i^m]$ are torsion classes and of order two. Since

$$\Sigma_i [M_i^{n-m}, f]_2 [V_i^m]_2 = 0,$$

it follows from (17.2) that $[V_i^m]_2 = 0$ and thus $[V_i^m] = 0$.

CHAPTER III
The G-bordism groups

This chapter serves as a transition between the preceding purely topological chapters and the following chapters which deal specifically with periodic maps. In section 19 we interpret $\Omega_n(B(G))$, G a finite group, as a group of bordism classes of pairs (G, M^n) consisting of a closed oriented manifold M^n and an orientation preserving differentiable free action of G on M^n. Section 20 deals with a bordism analogue of the classical transfer homomorphism. Sections 21 and 22 give elementary properties of differentiable actions on a compact n-manifold B^n, in particular an equivariant collaring theorem and a discussion of tubular neighborhoods of invariant submanifolds.

19. The principal G-bordism groups

We consider an equivariant bordism theory for principal G-manifolds, G a finite group. An *oriented G-manifold* is a pair consisting of a compact oriented manifold B^n together with an orientation preserving action of G on B^n as a group of diffeomorphisms; we shall denote such a pair by (G, B^n) or simply by B^n. A *principal oriented G-manifold* is an oriented G-manifold such that no element of the group other than the identity has a fixed point.

Two oriented G-manifolds (G, B_1^n) and (G, B_2^n) are *equivalent* provided there exists an orientation preserving equivariant diffeomorphism φ of B_1^n onto B_2^n; recall that φ is equivariant if $\varphi(gx) = g(\varphi(x))$ for all $g \in G$. We shall borrow the following result, to be proved in section 21. Let (G, B^n) denote a G-manifold. There is an open invariant set $U \supset \dot{B}^n$ and an equivariant diffeomorphism φ of $(G, B^n \times [0, 1))$ onto (G, U) with $\varphi(x, 0) = x$, where G acts on $B^n \times [0, 1)$ via $g(x, t) = (gx, t)$. This is just an analogue of (1.2).

Consider a closed oriented principal G-manifold (G, M^n). We say that (G, M^n) *equivariantly bords* if and only if there is a compact oriented principal G-manifold (G, B^{n+1}) with $(G, \dot{B}^{n+1}) = (G, M^n)$. Given (G, M_1^n) and (G, M_2^n), there is the usual disjoint union denoted by $(G, M_1^n \cup M_2^n)$. Let $-(G, M^n) = (G, -M^n)$. Define (G, M_1^n) to be *equivariantly bordant* to (G, M_2^n) if and only if the disjoint union $(G, M_1^n \cup -M_2^n)$ equivariantly bords.

Equivariant bordism is an equivalence relation on the closed oriented principal G-manifolds. To see that (G, M^n) is equivalent to itself we have only to consider $(G, M^n \times I)$, where $g(x, t) = (gx, t)$. Symmetry is obvious. Transitivity follows from the equivariant collaring theorem (21.2). Denote the equivariant bordism class of (G, M^n) by $[G, M^n]$, and the collection of all such classes by $\Omega_n(G)$. An abelian group structure is imposed on $\Omega_n(G)$ via disjoint union. We use the notation $\Omega_n(G)$ here by analogy with the notation $H_n(G; Z)$ for the homology groups of a finite group.

The weak direct sum $\Omega_*(G) = \Sigma_0^\infty \Omega_n(G)$ is a graded right Ω-module. Specifically, from a closed oriented principal G-manifold (G, M^n) and a closed oriented manifold V^m a closed oriented principal G-manifold $(G, M^n \times V^m)$ is given by $g(x, y) = (gx, y)$. We speak of $\Omega_*(G)$ as the *oriented bordism module* of the finite group G.

Let $E(G)$ be a universal space for the group G, in particular a CW complex upon which G acts freely and with every skeleton of $E = E(G)$ a finite CW complex. Also let $B(G) = E(G)/G$ denote the classifying space of G; we may also suppose it a CW complex with skeletons finite.

(19.1) Theorem. *For each closed oriented principal G-manifold, let $v: M^n \to M^n/G$ denote the orbit map. There exists a unique differentiable structure on M^n/G in which v is a local diffeomorphism, and a unique orientation for M^n/G such that v preserves orientation locally. There exists an equivariant map $f: M^n \to E(G)$, inducing a map $\bar{f}: M^n/G \to B(G)$. The correspondence $[G, M^n] \to [M^n/G, \bar{f}]$ is an isomorphism of the $\Omega_*(G)$ defined above onto the $\Omega_*(B(G))$ defined in Chapter I.*

Proof. To show the correspondence well-defined, suppose $(G, M^n) = (G, \dot{B}^{n+1})$. By universality of $E(G)$, $f: M^n \to E(G)$ can be extended to an equivariant map $F: B^{n+1} \to E(G)$, inducing an $\bar{F}: B^{n+1}/G \to B(G)$ which extends $\bar{f}: M^n/G \to B(G)$. Moreover, B^{n+1}/G is seen to be a compact oriented manifold whose boundary is M^n/G. Hence $[M^n/G, \bar{f}] = 0$ in $\Omega_n(B(G))$, and the correspondence is well-defined.

Suppose now that (V^n, f) is a closed oriented singular manifold in $B(G)$. There is the diagram

$$
\begin{array}{c}
E(G) \\
\downarrow{\scriptstyle v} \\
V^n \xrightarrow{\;f\;} B(G)
\end{array}
$$

inducing as usual a principal G-bundle over V^n. Let $M^n \subset V^n \times E$ be defined then as the set of (x, y) with $fx = vy$. Let G act on M^n via $g(x, y) = (x, gy)$. The projection $v: M^n \to M^n/G$ is a local homeomorphism. A differentiable structure is imposed on M^n so that v preserves orientation locally. We thus get a closed oriented principal G-manifold (G, M^n), and $[G, M^n]$ maps into $[V^n, f]$. Thus the correspondence is an epimorphism. In a similar fashion it is seen to be a monomorphism.

We can now apply the results of Chapters I and II to $\Omega_*(G)$. From section 7, we have the following.

(19.2) *For every finite group G there is a spectral sequence $\{E_{p,q}^r\}$ with $E_{p,q}^2 = H_p(G; \Omega_q)$ and with E^∞-term associated with a filtration of $\Omega_*(G)$.*

There is also an unoriented bordism module $\mathfrak{N}_*(G)$, defined by means of closed principal G-manifolds (G, M^n) in which there are no orientability requirements on M^n. From (17.2) we get the following.

(19.3) *For any finite group G the unoriented bordism module $\mathfrak{N}_*(G)$ is isomorphic to $H_*(G; Z_2) \otimes \mathfrak{N}$.*

There is also an analogue of the reduced bordism groups. Namely an augmentation $\varepsilon_* : \Omega_*(G) \to \Omega$ is defined by sending $[G, M^n]$ into $[M^n/G]$. It is seen that (19.1) identifies this with the augmentation $\Omega_*(B(G)) \to \Omega$. We can then define $\tilde{\Omega}_n(G) = \mathrm{Ker}\,[\varepsilon_* : \Omega_*(G) \to \Omega]$, and obtain such results as $\tilde{\Omega}_*(G) \cong \tilde{\Omega}_*(B(G))$ and $\Omega_n(G) = \Omega_n \oplus \tilde{\Omega}_n(G)$.

(19.4) **Theorem.** *If (G, M^n) is a closed oriented principal G-manifold where G is a finite group of order r, then $[M^n] = r[M^n/G]$.*

Proof. The quotient map $v : M^n \to M^n/G$ is a local diffeomorphism, and thus the bundle over M^n induced by v from the tangent bundle to M^n/G is the tangent bundle to M^n. Thus $v^* : H^*(M^n/G; Z) \to H^*(M^n; Z)$ maps the Pontryagin classes P_k of M^n/G onto the Pontryagin classes of M^n. Similarly $v^* : H^*(M^n/G; Z_2) \to H^*(M^n; Z_2)$ maps Stiefel-Whitney classes of M^n/G onto Stiefel-Whitney classes of M^n.

It is also seen that if σ_n, σ_n' denote the orientation classes of M^n and M^n/G respectively, then $v_*(\sigma_n) = r\sigma_n'$. Looking at Pontryagin numbers, then

$$\langle P_{i_1}(M^n) \ldots P_{i_k}(M^n), \sigma_n \rangle = \langle v^* P_{i_1}(M^n/G) \ldots v^* P_{i_k}(M^n/G), \sigma_n \rangle$$
$$= \langle P_{i_1}(M^n/G) \ldots P_{i_k}(M^n/G), v_*\sigma_n \rangle$$
$$= r\langle P_{i_1}(M^n/G) \ldots P_{i_k}(M^n/G), \sigma_n' \rangle .$$

Thus a Pontryagin number of M^n is r times the corresponding Pontryagin number of M^n/G. Precisely the same remark applies to Stiefel-Whitney numbers. Then $[M^n]$ and $r[M^n/T]$ have the same Pontryagin numbers and the same Stiefel-Whitney numbers, and the result follows.

Now (19.4) answers easily a question variants of which will arise in later chapters. Namely, which bordism classes of Ω_* admit representatives M^n upon which there exists a differentiable, orientation preserving free action of G? According to (19.4), every such bordism class is divisible by r. On the other hand, every bordism class $r[V^n]$ admits such representatives, namely r copies of V^n (that is, $G \times V^n$), permuted appropriately by G. Hence the class of such bordism classes is precisely the ideal $r\Omega$.

20. The transfer homomorphism

Let G be a finite group and $H \subset G$ a subgroup. Classically there is a homomorphism $i : H_n(H; Z) \to H_n(G; Z)$ and a transfer homomorphism $t : H_n(G; Z) \to H_n(H; Z)$. We shall investigate the corresponding homomorphisms relating $\Omega_*(G)$ and $\Omega_*(H)$.

First, let $A(G)$ denote the group of automorphisms of G and let $I(G) \subset A(G)$ be the normal subgroup of inner-automorphisms. We shall define an action of $A(G)/I(G)$ on $\Omega_*(G)$ as a group of Ω-module homomorphisms of degree 0. Let (G, M^n) be a closed oriented principal G-manifold and let $\gamma : G \to G$ be an automorphism. A new action $\gamma_*(G, M^n)$ is given by $g^*x = \gamma(g)(x)$. This is again an oriented principal G-space. We can set $\gamma_*([G, M^n]) = [\gamma_*(G, M^n)]$. This is an Ω-module automorphism of degree 0. Now suppose $\gamma(g) = hgh^{-1}$. Define $m : M^n \to M^n$ by $m(x) = hx$. Now $g^*(m(x)) = hgh^{-1}hx = hgx = m(gx)$, thus $m : (G, M^n) \to \to \gamma_*(G, M^n)$ is an orientation preserving equivariant diffeomorphism so $\gamma_*([G, M^n]) = [G, M^n]$. In this way we see that the quotient group $A(G)/I(G)$ acts on $\Omega_*(G)$.

Next consider $H \subset G$. We first define the transfer $t : \Omega_n(G) \to \Omega_n(H)$. Let (G, M^n) be a closed oriented principal G-manifold. This induces by restriction (H, M^n) which is still a closed oriented principal H-space. We simply let $t([G, M^n]) = [H, M^n]$. This defines an Ω-module homomorphism $t : \Omega_*(G) \to \Omega_*(H)$ with degree 0.

The homomorphism $i : \Omega_n(H) \to \Omega_n(G)$ is defined as follows. Begin with a closed oriented principal H-manifold, (H, M^n). We form the product $G \times M^n$ on which H acts by the rule $h(g, x) = (gh^{-1}, hx)$. Thus $G \times M^n$ is also a closed oriented principal H-manifold. We form the quotient $(G \times M^n)/H$. Let $((g, x))$ denote the point in the quotient corresponding to $(g, x) \in G \times M^n$. The group G acts on $(G \times M^n)/H$ by $\hat{g}((g, x)) = ((\hat{g}g, x))$. We set $i([H, M^n]) = [G, (G \times M^n)/H]$. Following the classical usage we refer to $t : \Omega_*(G) \to \Omega_*(H)$ as the transfer homomorphism. We should note here that the problem of computing the composite homomorphism $\Omega_n(G) \xrightarrow{t} \Omega_n(H) \xrightarrow{i} \Omega_n(G)$ is a complete mystery. This is quite unlike the classical situation for it is well known that $it : H_n(G; Z) \to H_n(G; Z)$ is multiplication by the index of H in G.

We shall determine $\Omega_n(H) \xrightarrow{i} \Omega_n(G) \xrightarrow{t} \Omega_n(H)$ under the assumption H is normal in G. This assumption is not really needed, but is suffices for our purposes. If H is normal in G there is a natural homomorphism $\tau : G/H \to A(H)/I(H)$ which assigns to $g \in G$ the automorphism $h \to ghg^{-1}$. We begin with (H, M^n) and we consider $[H, (G \times M^n)/H] \in \Omega_n(H)$. This will be $ti([H, M^n])$. We note that $h((g, x)) = ((hg, x)) = ((gg^{-1}hg, x)) = ((g, g^{-1}hgx))$. We shall use this formulation of the action of H on $(G \times M^n)/H$.

Let $\eta : G \to G/H$ denote the quotient homomorphism. We select $g_1, g_2, \ldots, g_k \in G$, one from each coset of H in G. First we define $(H, G/H \times M^n)$ by $g(\eta(g_j), x) = (\eta(g_j), g_j^{-1}hg_jx)$. Observe that for a fixed g_j, $(\eta(g_j) \times M^n)$ is H-invariant. Actually the action of H on $(\eta(g_j) \times M^n)$ is $\tau(\eta(g_j))_*(H, M^n)$. Thus $[H, G/H \times M^n] = \Sigma_1^k \tau(\eta(g_j))_*[H, M^n]$. We see how the action of $A(H)/I(H)$ on $\Omega_n(H)$ enters our picture.

We must show $[H, (G \times M^n)/H] = [H, G/H \times M^n]$. We define $m : (G \times M^n)/H \to G/H \times M^n$ as follows. Given $((g, x))$ there is a unique $j, h \in H$ for which $g = g_j h$. Let $m((g, x)) = (\eta(g_j), hx)$. Suppose $((g, x)) = ((\hat{g}, y))$, then $g\hat{h}^{-1} = g$, $\hat{h}x = y$. Thus $\hat{g} = g_j h \hat{h}^{-1}$, so $m((\hat{g}, y)) = (\eta(g_j), h\hat{h}^{-1}y) = (\eta(g_j), hx)$. In this way m is well defined. Now we see that

$$m\big(\hat{h}((g, x))\big) = m((\hat{h}g, x)) = m((g, g^{-1}\hat{h}gx)) = (\eta(g_j), hh^{-1}g_j^{-1}\hat{h}g_j hx)$$

$$= (\eta(g_j), g_j^{-1}\hat{h}g_j hx) = \hat{h}(\eta(g_j), hx) = \hat{h}m((g, x)) .$$

Therefore m is equivariant. The reader may show m is $1 - 1$ onto.

(20.1) Theorem. *Let $H \subset G$ be a normal subgroup, then* $ti([H, M^n]) = \Sigma_1^k \tau(\eta(g_j))_*([H, M^n])$.

The right side is clearly independent of our choice of $g_1, \ldots, g_k$.

(20.2) Corollary. *If H belongs to the center of G then* $ti(H, M^n]) = k[H, M^n]$ *where k is the index of H in G.*

We shall see that (20.1) corresponds to the case of the composition $ti : H_n(H, Z) \to H_n(H; Z)$. Our task is to now show that the diagram

$$
\begin{array}{ccccc}
\Omega_n(H) & \xrightarrow{i} & \Omega_n(G) & \xrightarrow{t} & \Omega_n(H) \\
\downarrow{\mu} & & \downarrow{\mu} & & \downarrow{\mu} \\
H_n(H;Z) & \xrightarrow{i} & H_n(G;Z) & \xrightarrow{t} & H_n(H;Z)
\end{array}
$$

is commutative.

Let (G, W) be the universal principal G-space with $W/G = B(G)$. Obviously $W/H = B(H)$. We may form the principal G-space $(G, (G \times W)/H)$. This admits a unique equivariant map $i:(G, (G \times W)/H) \to (G, W)$, which induces $\hat{i} : ((G \times W)/H)/G \to B(G)$. We identify $((G \times W)/H)/G$ with $W/H = B(H)$ to obtain $\hat{i} : B(H) \to B(G)$ which produces the commutative diagram

$$
\begin{array}{ccc}
\Omega_n(B(H)) & \xrightarrow{\hat{i}_*} & \Omega_n(B(G)) \\
\downarrow{\mu} & & \downarrow{\mu} \\
H_n(B(H);Z) & \xrightarrow{\hat{i}_*} & H_n(B(G);Z)
\end{array} \quad ,
$$

which is just

$$
\begin{array}{ccc}
\Omega_n(H) & \xrightarrow{i} & \Omega_n(G) \\
\downarrow{\mu} & & \downarrow{\mu} \\
H_n(H;Z) & \xrightarrow{i} & H_n(G;Z)
\end{array} \quad .
$$

Now choose a closed oriented principal G-manifold which is N-universal (G, V^m), $N \gg n$. We may take V^m to be an appropriate Stiefel manifold. We have $V^m/H \xrightarrow{\hat{i}} V^m/G$. Let $f : W^n \to V^m/G$ define $[W^n, f] \in$

$\in \Omega_n(V^m/G)$. If $m > 2n$ we may by the Whitney embedding theorem (10.2) replace $f: W^n \to V^m/G$ by an inclusion $W^n \subset V^m/G$. In fact, we may replace $\Omega_n(V^m/G)$ by $L_n(V^m/G)$, the group of L-equivalence classes of closed regular n-dimensional submanifolds in V^m/G. The local diffeomorphism $\hat{i}: V^m/H \to V^m/G$ is transverse regular on W^n, thus $\widetilde{W}^n = \hat{i}^{-1}(W^n) \subset V^m/H$ is a closed regular submanifold in V^m/H. In this manner a homomorphism $t: L_n(V^m/G) \to L_n(V^m/H)$ is defined. This is the transfer homomorphism. In homology the transfer is simply the Hopf umkehr homomorphism of $i: V^m/H \to V^m/G$. Obviously the diagram

$$
\begin{array}{ccc}
\Omega_n(G) & \xrightarrow{\,t\,} & \Omega_n(H) \\
\downarrow{\scriptstyle \mu} & & \downarrow{\scriptstyle \mu} \\
H_n(G;Z) & \xrightarrow{\,t\,} & H_n(H;Z)
\end{array}
$$

commutes.

(20.3) **Theorem.** *For any subgroup $H \subset G$ the diagram*

$$
\begin{array}{ccccc}
\Omega_n(H) & \xrightarrow{\,i\,} & \Omega_n(G) & \xrightarrow{\,t\,} & \Omega_n(H) \\
\downarrow{\scriptstyle \mu} & & \downarrow{\scriptstyle \mu} & & \downarrow{\scriptstyle \mu} \\
H_n(H;Z) & \xrightarrow{\,i\,} & H_n(G;Z) & \xrightarrow{\,t\,} & H_n(H;Z)
\end{array}
$$

is commutative.

We shall use the transfer homomorphism later as a computational tool in studying $\Omega_*(G)$.

21. The G-bordism groups

In section 19 we have defined the bordism groups of *principal* G-manifolds (G, M^n). Here we point out that there are similar G-bordism groups in which the actions are not necessarily free. In order to prove transitivity of the G-bordism relation, we first prove the equivariant form of the collaring theorem (1.2).

Let G be a compact Lie group. Let B^n be a compact manifold; fix once and for all a closed manifold M^n which contains B^n as a regular submanifold. The existence of M^n follows from (1.2); for example we can take M^n to be the double of B^n. A *differentiable action* of G on B^n is a map $\eta: G \times B^n \to B^n$ such that

    i) $\eta(e, x) = x$, for e the identity;

    ii) $\eta(g_1, \eta(g_2, x)) = \eta(g_1 g_2, x)$;

    iii) η is a differentiable map.

This last condition implies that each point $(g, x) \in G \times B^n$ there is a neighborhood in $G \times M^n$ to which η may be extended to a differentiable map. By a result of MILNOR [27] there exists an open neighborhood U

of B^n in M^n and a differentiable extension $\hat{\eta}: G \times U \to M^n$ of η. Fix U and $\hat{\eta}$ once and for all.

(21.1) *Suppose the compact Lie group G acts differentiably on the compact n-manifold B^n and also on the closed m-manifold V^m. If A is a closed invariant subset of B^n and if $f: A \to V^m$ is an equivariant differentiable map, then there is an open invariant set W, $A \subset W \subset B^n$, and an equivariant differentiable extension $F: W \to V^m$.*

Proof. Choose an open set W_1 in M^n with $A \subset W_1 \subset U$ and a differentiable extension $f: W_1 \to V^m$ of f. There exists an open set W_2 in M^n with $A \subset W_2 \subset W_1 \subset U$ and $\hat{\eta}(G \times W_2) \subset W_1$.

We use now the Mostow-Palais embedding theorem [31, 32]. There is an orthogonal representation of G on R^k and an equivariant differentiable embedding $\varphi: V^m \to R^k$. There is also an open invariant set $0 \supset \varphi(V^m)$ and an equivariant differentiable retraction $\varrho: 0 \to \varphi(V^m)$.

Denote by $\bar{f}(g, x)$ the composite $G \times W_2 \xrightarrow{\hat{\eta}} W_1 \xrightarrow{\varphi f} \varphi(V^m) \subset R^k$. Define $\bar{F}(x) = \int_G g^{-1}\bar{f}(g, x)\, dg$; then $\bar{F}$ is a differentiable map of W_2 into R^k and $\bar{F} = \varphi f$ on A. Choose an invariant set W open in B^n, $A \subset W \subset W_2$, with $\bar{F}(W) \subset 0$. The desired extension of $f: A \to V^m$ is $\varphi^{-1}\varrho\bar{F}: W \to V^m$.

There is the following corollary. *If G acts differentiably on the compact n-manifold B^n, there is an open invariant set W, $\dot{B}^n \subset W \subset B^n$, and an equivariant differentiable retraction $r: W \to \dot{B}^n$.*

We now prove the following. Here G acts on $\dot{B}^n \times [0, 1)$ via $g(x, t) = (gx, t)$.

(21.2) **Theorem.** *Suppose that G acts differentiably on the compact n-manifold B^n. There is an open invariant set V with $\dot{B}^n \subset V \subset B^n$ and an equivariant diffeomorphism $h: V \to \dot{B}^n \times [0, 1)$ with $h(x) = (x, 0)$ for $x \in \dot{B}^n$.*

Proof. Consider the tangent bundle $q: E \to B^n$, the restriction to B^n of the tangent bundle to M^n. The group G acts on E as a group of bundle maps, covering the action of G on B^n. For a real-valued differentiable function f and for a tangent vector $v \in B_x^n$ (the tangent space at x), denote by $\langle f, v \rangle$ the directional derivative of f in the direction v. Denoting by fg the composite function $f(gx)$, we have $\langle fg, v \rangle = \langle f, gv \rangle$. Let $F(x) = \int_G f(gx)\, dg$. Then $\langle F, v \rangle = \int_G \langle fg, v \rangle\, dg = \int_G \langle f, gv \rangle\, dg$.

Now let V_1 be an open neighborhood of $\dot{B}^n$ in B^n for which there is a diffeomorphism $h: V_1 \to \dot{B}^n \times [0, 1)$ with $h(x) = (x, 0)$, $x \in \dot{B}^n$ (see (1.2)). Let $f: B^n \to R$ be a differentiable map which on V_1 is the projection of $h(x)$ into $[0, 1)$. At a point $x \in \dot{B}^n$, if v is a tangent vector pointing toward the interior of B^n then $\langle f, v \rangle > 0$. Moreover $\langle f, v \rangle = 0$ if and only if v is tangent to $\dot{B}^n$. Let $F(x) = \int_G f(gx)\, dg$. Then for $x \in \dot{B}^n$ and v a tangent vector at x pointing toward the interior of B^n,

$$\langle F, v \rangle = \int_G \langle f, gv \rangle\, dg \geqq 0.$$

Moreover, $\langle F, v \rangle = 0$ if and only if v is tangent to $\dot{B}^n$.

Select an open invariant W, $\dot{B}^n \subset W \subset V_1$, for which there is an equivariant differentiable retraction $r: W \to \dot{B}^n$. Then define $h: W \to \dot{B}^n \times [0, \infty)$ by $h(x) = (r(x), F(x))$. Along $\dot{B}^n$, h induces an isomorphism of the tangent spaces; for $x \in \dot{B}^n$, $h(x) = (x, 0)$ and h is a diffeomorphism of $\dot{B}^n$ onto $\dot{B}^n \times 0$. There is then an open invariant W_1, $\dot{B}^n \subset W_1 \subset W$, such that $h: W_1 \to \dot{B}^n \times [0, 1)$ is an equivariant diffeomorphism onto an open subset. There is an $\varepsilon > 0$ for which $\dot{B}^n \times [0, \varepsilon)$ lies in the image of h. Let $V = h^{-1}(\dot{B}^n \times [0, \varepsilon))$ and define $h': V \to \dot{B}^n \times [0, 1)$ by $h(x) = (r(x), (1/\varepsilon) F(x))$. The theorem then follows.

Suppose now that (G, B_1^n) and (G, B_2^n) are compact G-manifolds; that is, pairs consisting of a compact manifold B_i^n and a differentiable action of G on B_i^n. If the boundaries $(G, \dot{B}_1^n)$ and $(G, \dot{B}_2^n)$ are joined by an equivariant diffeomorphism φ, then we can use (21.2) to sew B_1^n and B_2^n together along their boundaries. That is, by identifying B_1^n and B_2^n along their boundary via φ to obtain B^n, we obtain a compact G-manifold (G, B^n).

As in section 19, a closed oriented G-manifold (G, M^n) consists of a closed oriented manifold and an orientation preserving differentiable action of the compact Lie group G on M^n. The *isotropy group* G_x of $x \in M^n$ is the subgroup $\{g : gx = x\}$ of G; it is clear that $G_{gx} = g G_x g^{-1}$. Fix a non-empty collection A of subgroups of G such that if $K \in A$ then $gKg^{-1} \in A$ for any $g \in G$. A G-manifold (G, M^n) is A-free if each isotropy group G_x is contained in an element of A.

We can now define an Ω-module $\Omega_*(G, A)$, generalizing $\Omega_*(G)$ in that free actions are replaced by A-free actions. To define $\Omega_*(G, A)$, we simply define a bordism relation on the closed oriented A-free G-manifolds duplicating that of section 19. In particular if A is the collection $S(G)$ of all subgroups of G, we get an *unrestricted G-bordism module*. If A consists of the single subgroup $\{e\}$, we get $\Omega_*(G)$. Similarly we can define unoriented G-bordism groups $\mathfrak{N}_*(G, A)$. It will be seen in the remainder of this work that $\Omega_*(G, A)$ is computed in very few cases. Nevertheless, its existence will be quite useful.

22. Tubular neighborhoods

In section 10, we have defined the classical tubular neighborhoods. Here we extend that discussion a bit to cover closed G-manifolds (G, M^n). We go on to a preliminary discussion of the stationary point structure of a closed G-manifold (G, M^n). This consists of the set F of stationary points, a disjoint finite union of i-manifolds F^i, $0 \leq i \leq n$, of the normal bundles to the F^i, and of the induced action of G on the normal bundles.

Let (G, M^n) be a closed G-manifold. By the usual averaging process, there is a Riemannian metric on M^n with respect to which G is a group of isometries. Let V^m be a closed submanifold of M^n, regularly embedded and invariant under the action of G. As in section 10, there is the tubular neighborhood N of V^m, of radius ε for each ε sufficiently small. Now N is always a regularly embedded submanifold of M^n. Since V^m is invariant and G is a group of isometries, it is also seen that N is invariant under the action of G. If $F_1, \ldots, F_k$ is a pairwise disjoint collection of closed invariant submanifolds of M^n, with dimensions varying with i, then a tubular neighborhood of $F = \cup F_i$ is by definition a disjoint collection of tubular neighborhoods of the various F_i.

Summarizing the above, let $\xi : E \to V^m$ denote the normal cell-bundle to V^m in M^n. Then E is a compact n-manifold; moreover G acts differentiably on E via the differentials $dg : E \to E$, and we may consider $V^m \subset E$. According to the above there is an equivariant diffeomorphism h of (G, E) onto (G, N), and h may be considered the identity on V^m.

If $x \in M^n$ is a stationary point, then x can be considered a 0-manifold so we may apply the above. In this case the normal bundle is the space M_x of tangent vectors to M^n at x, and we have an orthogonal representation of G on M_x. By the diffeomorphism h we may map the unit cell of M_x via an equivariant diffeomorphism onto invariant neighborhoods of x in M^n. This yields the linearization of the action of G at a stationary point [29, p. 206]. It follows immediately that the component F containing x of the set of stationary points is a regular submanifold of M^n. Moreover the tangent space to F at x is the subspace of the tangent space to M^n at x which is pointwise fixed under the linear representation of G on M_x.

Suppose now that F is the set of stationary points of (G, M^n). Let F^i denote the union of the i-dimensional components of F. Then F^i is a closed regularly embedded submanifold of M^n, invariant under G. There is the normal cell bundle $\xi : E_i \to F^i$, and the action of G on E_i; the action on each fiber is an orthogonal representation of G. Denote by E the union $\cup E_i$, and by $\xi : E \to F$ the map $\cup E_i \to \cup F^i$. We shall occasionally speak of $\xi : E \to F$ as a cell bundle. We can now identify (G, E) with the tubular neighborhood (G, N) of F in M^n. A particularly simple case is that in which $G = Z_2$. Then Z_2 acts orthogonally on the fibers of ξ, and leaves only the zero vectors fixed. The only such action of Z_2 on a vector space is via the antipodal map.

(22.1) **Theorem.** *Suppose that G acts differentiably on the closed oriented manifolds M_1^n and M_2^n, preserving the orientation. Suppose that the sets F_1 and F_2 of stationary points have tubular neighborhoods which are equivariantly diffeomorphic via an orientation preserving diffeomorphism φ. There exists a closed oriented manifold V^n and a differentiable action of G on V^n without stationary points, with*

$$[M_1^n] - [M_2^n] = [V^n] \, .$$

Proof. Denote by N_1 and N_2 the tubular neighborhoods. Form $W^{n+1} = (-M_1^n) \times I \cup M_2^n \times I$. Form the identification space W_1^{n+1} of W^{n+1}, where $(x, 1) \in N_1 \times 1$ is identified with $(\varphi(x), 1) \in N_2 \times 1$ for all $x \in N_1$. The angle straightening technique may be used to show that W_1^{n+1} is a differentiable manifold. The boundary of W_1^{n+1} consists of the disjoint union of M_1, $-M_2$ and a third manifold V^n described as follows; V^n is formed from $-(M_1^n \backslash \mathrm{Int}\, N_1)$ and $M_2^n \backslash \mathrm{Int}\, N_2$ by identifying their boundary via φ. By (21.2), we see that the differentiable structure can be put on V^n so that G acts differentiably. Moreover, G acts on V^n without stationary points. The theorem follows.

There is an unoriented version of (22.1).

CHAPTER IV
Differentiable Involutions

We now begin our major undertaking, the study of differentiable periodic maps. Most of our attention is given to maps of prime period p. There appear to be two cases to treat, $p = 2$, and p odd. In this chapter we treat the case $p = 2$; that is, we deal with differentiable involutions T on closed manifolds M^n. A distinctive feature of the case $p = 2$ is that we may ignore matters of orientation.

In section 23, we give the complete structure of the bordism module $\mathfrak{N}_*(Z_2)$ of fixed point free involutions (T, M^n). In the succeeding sections we study the fixed point sets of involutions $T : M^n \to M^n$. The results typically relate the bordism class $[M^n]_2$ to the fixed point set and its normal bundle. The concluding section 28 gives the structure of the bordism theory of all differentiable involutions (T, M^n).

23. Fixed point free involutions

We begin our study of involutions with a further analysis of $\mathfrak{N}_*(Z_2)$ (see section 19). We go on to introduce an important class of fixed point free involutions, the bundle involutions associated with sphere bundles.

An *involution* is a homeomorphism $T : X \to X$ of period 2; the involutions are identified with the actions of the group Z_2 and will be denoted by (T, X). The differentiable fixed point free involutions (T, M^n) on closed manifolds are the closed principal Z_2-manifolds of section 19. There is then the bordism relation of section 19 and the bordism group $\mathfrak{N}_n(Z_2)$. Specifically (T, M^n) bords, written $[T, M^n]_2 = 0$, if there is a differentiable fixed point free involution (S, B^{n+1}) on a compact manifold with $\dot{B}^{n+1} = M^n$ and $S \,|\, M^n = T$.

Given a differentiable fixed point free involution (T, M^n), there is an equivariant map $M^n \to E(Z_2)$, where $E(Z_2)$ is a universal Z_2-bundle, and an induced map $\varrho : M^n/T \to B(Z_2)$ unique up to homotopy. Now

$H^*(B(Z_2); Z_2)$ is a polynomial algebra with a single one-dimensional generator denoted here by c. There is then $\varrho^*(c) \in H^1(M^n/T; Z_2)$; we denote $\varrho^*(c)$ by c and call it the *characteristic class of the involution*. It is of course the Whitney class of the Z_2-bundle $M^n \to M^n/T$.

We obtain the following directly from (17.2).

(23.1) Theorem. *The bordism class $[T, M^n]_2$ in $\mathfrak{N}_n(Z_2)$ of a differentiable fixed point free involution on a closed manifold is uniquely determined by the integers mod two $\langle w_{i_1} \ldots w_{i_k} c^m, \sigma_n \rangle$ where the w_i are the Stiefel-Whitney classes of M^n/T, $\sigma_n \in H_n(M^n/T; Z_2)$ is the fundamental class of M^n/T, and $i_1 + \cdots + i_k = n - m$.*

The above numbers $\langle w_{i_1} \ldots w_{i_k} c^m, \sigma_n \rangle$ will be called the *involution numbers*.

It follows from (8.3) that the $\mathfrak{N}$-module $\mathfrak{N}_*(Z_2)$ is a free $\mathfrak{N}$-module with one base element in each dimension.

(23.2) Theorem. *Suppose for each $n = 0, 1 \ldots$ we have a differentiable fixed point free involution (T, X^n) on a closed manifold, such that for each n the involution number $\langle c^n, \sigma_n \rangle$ of (T, X^n) is non-zero. Then $\{[T, X^n]_2\}$ is a basis for the $\mathfrak{N}$-module $\mathfrak{N}_*(Z_2)$. In particular the antipodal map (A, S^n) of the n-sphere has $\langle c^n, \sigma_n \rangle \neq 0$, and $\{[A, S^n]_2\}$ is a base.*

Proof. Considering the map $\varrho : X^n/T \to B(Z_2)$, we have

$$\langle c^n, \varrho_* \sigma_n \rangle = \langle \varrho^* c^n, \sigma_n \rangle \neq 0 .$$

Hence $\varrho_*(\sigma_n) \neq 0$ in $H_n(B(Z_2); Z_2)$. The above now follows from (8.3).

Note that (23.2) implies that every fixed point free (T, M^n) has a unique representation

$$[T, M^n]_2 = \Sigma_{m=0}^n [A, S^{n-m}]_2 [V^m]_2 .$$

We turn now to the bundle involutions. Consider a differentiable sphere bundle $q : B \to V^n$ over the closed manifold V^n; the fiber is the $(k-1)$-sphere S^{k-1} and the structural group is $0(k)$. Note that B is a closed $(n+k-1)$-manifold. The antipodal map $A : S^{k-1} \to S^{k-1}$ lies in the center of $0(k)$. Hence it is seen that there is a fiber preserving differentiable fixed point free involution (T, B) which on each fiber reduces to the antipodal map. We refer to (T, B) as the *bundle involution*; this involution was studied in [14].

Consider the diagram

$$B \longrightarrow B/T$$
$$q \searrow \quad \swarrow p$$
$$V^n$$

We see that $p : B/T \to V^n$ is a bundle with real projective space P_{k-1} as fiber. If $P_{k-1} \subset B/T$ is a fiber then by the naturality of Whitney classes the image of $c \in H^1(B/T; Z_2)$ under $i^* : H^1(B/T; Z_2) \to H^1(P_{k-1}; Z_2)$ is the generator of $H^*(P_{k-1}; Z_2)$. The fiber P_{k-1} is thus totally non-homologous to zero in B/T. By the Leray-Hirsch theorem [5, p. 129] the mod 2 spectral sequence of p collapses. It follows that for any class $h^r \in H^r(B/T; Z_2)$ there are unique classes $e_r, e_{r-1}, \ldots, e_{r-k+1}$ in $H^*(V^n; Z_2)$ for which

$$h^r = p^*(e_r) + p^*(e_{r-1})c + \cdots + p^*(e_{r-k+1})c^{k-1} .$$

Consider now the tangent bundle ξ of B/T. The differentiable fiber map p splits ξ into a Whitney sum $\xi_1 \oplus \xi_2$ [7, p. 482]. Here ξ_1 is the normal bundle to the fiber and ξ_2 is the tangent bundle to the fiber. By naturality, the total Whitney class of ξ_1 is $1 + p^*(w_1) + \cdots + p^*(w_n)$, where the w_j are the Stiefel-Whitney classes of V^n. The total Whitney class of ξ_2 was computed by BOREL-HIRZEBRUCH [7, p. 517]; a proof will be given in section 31.

(23.3) **Borel-Hirzebruch.** *The total Whitney class of the tangent bundle ξ_2 along the fiber in B/T is given by $(1 + c)^k + (1 + c)^{k-1}p^*(v_1) + \cdots + p^*(v_k)$, where the v_i are the Whitney classes of the $0(k)$-bundle $q:B \to V^n$. Since ξ_2 is an $0(k-1)$-bundle it also follows that*

$$c^k = c^{k-1}p^*(v_1) + \cdots + p^*(v_k) .$$

Thus the total Stiefel-Whitney class of B/T is $(\Sigma_0^n p^*(w_j)) \times \times (\Sigma_0^k (1 + c)^{k-j} p^*(v_j))$. Explicitly,

$$(23.4) \qquad W_m = \Sigma_{p+q+r=m} \binom{k-p}{q} p^*(w_r v_p) c^q .$$

In principal this will enable us to compute the involution numbers of the bundle involution (T, B), although in practice we find it very difficult.

24. Fixed Point Sets of Differentiable Involutions

We come to the first installment of our study of fixed point sets F of differentiable involutions $T: M^n \to M^n$. We prove that F and its normal bundle determine in an explicit way the mod 2 bordism class $[M^n]_2$. Our first illustration of the interest of this fact is a geometric proof of a theorem of WALL that if M^n is a closed manifold then $M^n \times M^n$ is bordant mod 2 to an orientable manifold. A second illustration is the study of fixed point sets of conjugations of almost complex manifolds.

Consider a differentiable involution (T, M^n) on a closed manifold. Fix once and for all a Riemannian metric on M^n with respect to which T is an isometry. Denote by F the fixed point set of T, and let F^m, $0 \leq m \leq n$, denote the union of the m-dimensional components of the fixed point set. According to section 21, F^m is a regularly embedded submanifold of M^n. Note also that F^n consists of the components of M^n which are pointwise fixed under T.

There is now the normal bundle $E_m \to F^m$, $m < n$, to F^m in M^n, a differentiable $(n-m)$-vector space bundle with inner product, and the corresponding differentiable $(n-m-1)$-sphere bundle $q: B_m \to F^m$. For each $m < n$ there is the bundle involution (T, B_m) of the sphere bundle B_m. Denote the disjoint union $\cup_{m<n}(T, B_m)$ by (T, B); call B the *normal sphere bundle to the fixed point set*.

(24.1) *If (T, M^n) is a differentiable involution of a closed manifold, and if (T, B) is the bundle involution of the normal sphere bundle B to the fixed point set, then $[T, B]_2 = 0$ in $\mathfrak{N}_{n-1}(Z_2)$.*

Proof. We may as well suppose $F^n = \emptyset$. Let N be a tubular neighborhood of F (see section 21). Then $B^n = M^n \backslash \mathrm{Int}\, N$ is a regularly embedded invariant submanifold with boundary, on which T has no fixed points. Then $[T, \dot{B}^n]_2 = [T, B]_2 = 0$. The result follows.

We go now to the chief result of the section.

(24.2) **Theorem.** *Let (T, M^n) denote a differentiable involution on a closed manifold. Denote by $q: E \to F$ the normal bundle to the fixed point set F, by $q': E' \to F$ the Whitney sum of q with a trivial line bundle, and by $B' \to F$ the associated sphere bundle to q'. Denote by (T', B') the bundle involution of B'. Then $[M^n]_2 = [B'/T']_2$.*

Proof. We have first to point out that the definition of B' should be carried out in each dimension. That is, $E'_m \to F^m$ is the Whitney sum of $E_m \to F^m$ with a trivial line bundle (for $m = n$, $E'_n \to F^n$ is a trivial line bundle). Moreover $B'_m \to F^m$ is the associated sphere bundle, and $B' = \cup B'_m$.

The proof is obtained by considering the involutions $(T_1, M^n \times I)$ and $(T_2, M^n \times I)$, where

$$T_1(x, t) = (x, 1-t),\quad T_2(x, t) = (T(x), 1-t).$$

The fixed point set of T_1 is $M^n \times 1/2$, and the normal bundle to the fixed point set is a trivial line bundle. The fixed point set of T_2 is $F \times 1/2$; identify $F \times 1/2$ with F. The normal bundle to the fixed point set of T_2 is clearly the bundle $E' \to F$.

Define an equivariant diffeomorphism $\varphi: (T_1, M^n \times I) \cong (T_2, M^n \times I)$ by $\varphi(x, 1) = (T(x), 1)$, $\varphi(x, 0) = (x, 0)$. We adjoin $(T_1, M^n \times I)$ to $(T_2, M^n \times I)$ along their boundaries via φ to obtain an involution (T_3, M^{n+1}) on a closed manifold.

All that remains is to apply (24.1) to (T_3, M^{n+1}) to obtain $[A, S^0]_2 [M^n]_2 + [T', B']_2 = 0$. Thus $[A, S^0]_2 [M^n]_2 = [T', B']_2$. We pass to quotient spaces and obtain $[M^n]_2 = [B'/T']_2$.

The existence of such a result was suggested by the easy observation that a closed manifold which carries a fixed point free involution always bords mod 2. A simple geometric argument for this is given by noting that the mapping cylinder of the orbit map $M^n \to M^n/T$ is a compact manifold with boundary M^n. Now (24.2) is a generalization of this to the case in which fixed points are present. We now turn to elementary applications, first reproving the following result [42].

(24.3) **Wall.** *If M^n is a closed manifold, then $M^n \times M^n$ is bordant* mod 2 *to a closed orientable manifold.*

Proof. First consider the case in which M^n is of odd dimension $n = 2k + 1$. On the product $M^{2k+1} \times M^{2k+1}$ define T by $T(x, y) = (y, x)$. The fixed point set of the involution T is the diagonal $M^{2k+1} = \Delta \subset$ $\subset M^{2k+1} \times M^{2k+1}$. We apply (24.2) to T.

The normal bundle to Δ in $M^{2k+1} \times M^{2k+1}$ is equivalent to the tangent bundle of M^{2k+1}. If $B' \to M^{2k+1}$ is the Whitney join of the tangent sphere bundle to M^{2k+1} with a trivial 0-sphere bundle, then $[M^{2k+1} \times M^{2k+1}]_2 = [B'/T']_2$. We now show that B'/T' is orientable. The Whitney classes of $B' \to M^{2k+1}$ are the Stiefel-Whitney classes of M^{2k+1}. By (23.4), the Stiefel-Whitney class W_1 of B'/T' is

$$W_1 = p^*(w_1) + p^*(w_1) + \binom{2k+2}{1} c = 0.$$

Hence B'/T' is orientable and the result holds for n odd.

Next consider the involution $(T, P_n(C))$ given in by homogeneous coordinates by $T([z_1, \ldots, z_n]) = [\bar{z}_1, \ldots, \bar{z}_n]$. The fixed point set is precisely real projective space $P_n \subset P_n(C)$. We shall see that the normal bundle to P_n in $P_n(C)$ is equivalent to the tangent bundle of P_n. At a point in P_n consider the complex tangent space to $P_n(C)$. The real vectors are identified with the tangent space to P_n at this point, and the purely imaginary vectors make up the normal space. Now multiplication by $\sqrt{-1}$ will interchange the normal and tangent spaces, which provides the equivalence. We thus see by (24.2) that $[P_n(C)]_2$ is determined by the tangent bundle to P_n. But we can also consider $P_n \times P_n$ and its involution $(x, y) \to (y, x)$ to conclude that $[P_n \times P_n]_2$ is determined in the same way by the tangent bundle to P_n. Thus $[P_n(C)]_2 = [P_n \times P_n]_2$. This is of course a well-known fact [42].

We come now to the general case. Now $\mathfrak{N}$ is a polynomial algebra whose even dimensional generators can all be taken to be P_{2k}. Since $\mathfrak{N}$ is a polynomial algebra over Z_2, it is sufficient to check the theorem on the generators. But this follows from the preceding cases. The assertion follows.

The above discussion of the conjugation involution $(T, P_n(C))$ suggests a generalization. In $P_n(C)$ let V^m be a closed regular projective subvariety which is T-invariant; that is, $T(V^m) = V^m$. Such a variety in algebraic geometry is a real algebraic variety. The set $F \subset V^m$ of fixed points of (T, V^m) is called the real fold of V^m. We shall show that $[V^m]_2 = [F \times F]_2$. As far as we know this is a new fact about real folds.

We shall put the question in a more general setting. Let $\xi : E \to M^{2n}$ be the tangent bundle of an even dimensional manifold. An almost complex structure (M^{2n}, J) is determined by selecting a fiber preserving bundle isomorphism

$$
\begin{array}{ccc}
E & \xrightarrow{\;J\;} & E \\
& \xi \searrow \quad \swarrow \xi & \\
& M^{2n} &
\end{array}
$$

such that $J^2 = -I$ [7, p. 480]. A differentiable involution T on M^{2n} induces a bundle isomorphism T_* for which

$$
\begin{array}{ccc}
E & \xrightarrow{\;T_*\;} & E \\
\xi \downarrow & & \downarrow \xi \\
M^n & \xrightarrow{\;T\;} & M^n
\end{array}
$$

s a commutative diagram. We shall say that (T, M^{2n}) is a *conjugation* of the almost complex structure (M^{2n}, J) if and only if at each point $x \in M^{2n}$ the diagram

$$
\begin{array}{ccc}
E_x & \xrightarrow{\;T_*\;} & E_{T(x)} \\
J \downarrow & & \downarrow J \\
E_x & \xrightarrow{\;T_*\;} & E_{T(x)}
\end{array}
$$

anti-commutes; that is, $J T_* = - T_* J$. The idea is that T carries J into the conjugate almost complex structure given by $-J$.

(24.4) Theorem. *If (T, M^{2n}) is a conjugation of an almost complex structure on a closed manifold and if F is the fixed point set of T, then F is an n-dimensional submanifold and $[M^{2n}]_2 = [F \times F]_2$.*

Proof. This is valid if there are no fixed points since $[M^{2n}]_2 = 0$ in this case. Suppose that $F \neq \emptyset$, and let $x \in F$. Split E_x into I_x and N_x where

$$
I_x = \{v : T_*(v) = v\}, \; N_x = \{v : T_*(v) = -v\} .
$$

For any $v \in E_x$, $v = (v + T_*(v))/2 + (v - T^*(v))/2$, thus $E_x = I_x \oplus N_x$. Now I_x is just the subspace of vectors tangent to the fixed point set at x, while N_x is the normal space. Since $J T_* = - T_* J$ we have $J(I_x) = N_x$, $J(N_x) = I_x$. Thus $\dim I_x = n$. Hence the fixed point set F is of dimension n. Furthermore, J gives an equivalence between the normal and tangent bundles to the fixed point set. We may then use (24.2) just as in the proof of (24.3) to show the theorem.

We shall briefly indicate a natural way in which conjugations arise in connection with complex analytic manifolds. Let V^n be a closed complex analytic manifold, and let $\bar{V}^n$ be the conjugate complex structure.

On $V^n \times \overline{V}^n$ consider $T(x, y) = (y, x)$. This involution is a conjugation on the product manifold. A closed regular analytic manifold $M^m \subset V^n \times \overline{V}^n$ which is T-invariant may be thought of as a conjugation of the complex analytic structure on V^n. Now $F = \Delta \cap M^n$ is the real fold of this conjugation and $[M^m]_2 = ([F]_2)^2$.

25. The normal bundle and the tangent bundle to the fixed point set

In this section we give some evidence that if (T, M^n) is a differentiable involution for which $[M^n]_2 \neq 0$, then both the tangent bundle and the normal bundle to the fixed point set are somewhat complicated. The only new technique added to the preceding two sections is a bordism interpretation of the bundle involution.

Consider $\mathfrak{N}_*(BO(k))$. An element of $\mathfrak{N}_n(BO(k))$ is defined by a map $f : V^n \to BO(k)$ where V^n is a closed manifold. If two maps $V^n \to BO(k)$ are homotopic they represent the same bordism class. We may then think of a bordism class as given by a closed manifold V^n and a preferred homotopy class of maps $V^n \to BO(k)$. But the homotopy classes of maps $V^n \to BO(k)$ are in one-to-one correspondence with the vector space bundles over V^n. Thus we receive a bundle interpretation for $\mathfrak{N}_n(BO(k))$. Elements are represented by k-dimensional vector space bundles $\xi : E \to V^n$ over closed manifolds V^n; denote the bordism class by $[\xi]_2$ or $[\xi : E \to V^n]_2$. The bundle $\xi : E \to V^n$ bords if there is a bundle $\xi' : E' \to B^{n+1}$ with B^{n+1} a compact $(n + 1)$-manifold with $\dot{B}^{n+1} = V^n$ and with ξ the bundle induced on $V^n \subset B^{n+1}$. It is a noteworthy consequence of (17.2) that if the bundles $\xi_1 : E_1 \to V^n$ and $\xi_2 : E_2 \to V^n$ have the same Whitney classes, then they bord as bundles.

If we wish to use differentiable bundles in the above, we use for $BO(k)$ the Grassman manifold $M_{k, N}$ of unoriented k-planes through the origin in R^{k+N}. We take $N > n$ and use the unoriented differentiable bordism group $D_n(M_{k, N})$ as defined in section 9. In this fashion we can identify $\mathfrak{N}_n(BO(k))$ with bordism classes of differentiable k-plane bundles.

To each differentiable k-dimensional vector space bundle $\xi : E \to V^n$ we have the bundle involution (T, B) of section 23. The assignment $[\xi]_2 \to [T, B]_2$ gives a well-defined function $J : \mathfrak{N}_n(BO(k)) \to \mathfrak{N}_{n+k-1}(Z_2)$. It may be verified that J is an $\mathfrak{N}$-module homomorphism of degree $k - 1$. As we shall see, the homomorphism J is of central importance in the study of the fixed point set.

(25.1) Theorem. *Suppose that (T, M^n) is a differentiable involution on a closed manifold, and that F^m is the union of the m-dimensional components of the fixed point set of T. If the Whitney classes of the normal bundle to F^m are trivial, all $0 \leq m \leq n - 1$, then $[F^m]_2 = 0$ for $0 \leq m \leq \leq n - 1$ and $[M^n]_2 = [F^n]_2$.*

Proof. Consider the normal bundle $\xi_m : E_m \to F^m$, $0 \leq m \leq n-1$. Since the Whitney classes of ξ are trivial, it follows from (17.2) that ξ_m is bordant in $\mathfrak{N}_m(BO(n-m))$ to the trivial bundle $\eta_m : E^* \to F^m$. Thus $J([\xi_m]_2) = J([\eta_m]_2) = [A, S^{n-m-1}]_2 [F^m]_2$. Hence

$$0 = [T, B]_2 = \Sigma_0^{n-1} [T, B_m]_2 = \Sigma_0^{n-1} J([\xi_m]_2) = \Sigma_0^{n-1} [A, S^{n-m-1}]_2 [F^m]_2.$$

Since $\{[A, S^k]_2\}$ is a base for $\mathfrak{N}_*(Z_2)$, then $[F^m]_2 = 0$ for $0 \leq m \leq n-1$.

Consider now the bundle $\xi' : E' \to F$ of (24.2). The Whitney classes of ξ' are 0, thus as above

$$[T', B']_2 = \Sigma_0^n [A, S^{n-m}]_2 [F^m]_2 = [A, S^0]_2 [F^n]_2$$

and hence by (24.2), $[A, S^0]_2 [M^n]_2 = [A, S^0]_2 [F^n]_2$ and $[M^n]_2 = [F^n]_2$.

The above theorem was arrived at as a generalization of the following: *if (T, M^n) is a differentiable involution on a closed manifold for $n > 0$, then T cannot have precisely an odd number of fixed points.* We now go on to a sort of dual version of (25.1).

(25.2) Theorem. *Let (T, M^n) be a differentiable involution on a closed n-manifold, and let the fixed point set F^k of T be a connected k-manifold. If all the Stiefel-Whitney classes of F^k vanish, then $[M^n]_2 = 0$.*

Proof. As in (24.2), we consider the normal bundle $\xi : E \to F^k$ to F^k and the normal $(n-k-1)$-sphere bundle $q : B \to F^k$. There is the bundle involution (T, B) with $[T, B]_2 = J([\xi]_2) = 0$, and the projective space bundle $p : B/T \to F^k$. Suppose now we could prove that $[\xi]_2 = 0$; that is, ξ bords as a bundle. It follows easily that the Whitney sum ξ' of ξ with a trivial line bundle also bords. Then $J([\xi']_2) = 0$. We then have from (24.2)

$$[A, S^0]_2 [M^n]_2 = [T', B']_2 = J([\xi']_2) = 0$$

and $[M^n]_2 = 0$.

To complete the theorem, it is thus sufficient to prove the following lemma.

(25.3) Lemma. *Suppose $\xi : E \to V^m$ is a k-dimensional vector space bundle over the connected manifold V^m, and that all the Stiefel-Whitney classes of V^m are trivial. Then $J([\xi]_2) = 0$ in $\mathfrak{N}_{m+k-1}(Z_2)$ if and only if $[\xi]_2 = 0$ in $\mathfrak{N}_m(BO(k))$.*

Proof. Denote by (T, B) the bundle involution associated with ξ, and by $p : B/T \to V^m$ the associated fiber map. If $J([\xi]_2) = 0$ then $[T, B]_2 = 0$, and hence the involution numbers of (T, B) all vanish. We shall show for any partition $r = i_1 + \cdots + i_j$ that $p^*(v_{i_1} \ldots v_{i_j}) \times c^{m+k-1-r} = 0$ in $H^*(B/T; Z_2)$. Here the v are the Whitney classes of ξ. For $r = 0$, $c^{m+k-1} = 0$, since $\langle c^{m+k-1}, \sigma_{m+k-1} \rangle = 0$ and B/T is connected.

Suppose the remark has been shown for $r < r_0$. Choose a partition $r_0 = i_1 + \cdots + i_j$; then $W_{i_1} \ldots W_{i_j} c^{m+k-1-r_0} = 0$ where the W_i are the Stiefel-Whitney classes of B/T. From (23.4), $W_i = p^*(v_i) +$ terms involving c and hence $W_{i_1} \ldots W_{i_j} c^{m+k-1-r_0} = p^*(v_{i_1} \ldots v_{i_j}) c^{m+k-1-r_0}$ plus terms of higher power in c. We may employ the inductive hypothesis to eliminate the higher order terms. Thus $p^*(v_{i_1} \ldots v_{i_j}) c^{m+k-1-r_0} = 0$. Letting $r_0 = m$, we get $p^*(v_{i_1} \ldots v_{i_j}) c^{k-1} = 0$ for every partition of m. Hence $v_{i_1} \ldots v_{i_j} = 0$. All Whitney numbers of ξ are then seen to be zero by (17.2). Thus $[\xi]_2 = 0$. Clearly $[\xi]_2 = 0$ implies $J([\xi]_2) = 0$ and the lemma follows.

An extension of the above shows that (25.2) holds if F^k is required to be a k-manifold, but is not necessarily connected. The conclusion is false if the components of F are allowed to be of different dimensions. For example, there is an involution on P_2 whose fixed point set consists of a point and a simple closed curve.

26. The Smith homomorphism

In this section we set up some techniques needed in the following sections. The most important of these is a homomorphism $A : \mathfrak{N}_n(Z_2) \to \mathfrak{N}_{n-1}(Z_2)$ which we call the Smith homomorphism. We include its definition in the following theorem.

(26.1) **Theorem.** *Suppose (T, M^n) is a differentiable fixed point free involution on a closed manifold. For $N \geq n$ there exists a differentiable equivariant map $g : (T, M^n) \to (A, S^n)$ which is transverse regular on $S^{N-1} \subset S^N$. Let $V^{n-1} = g^{-1}(S^{N-1})$. The function $\Delta : \mathfrak{N}_n(Z_2) \to \mathfrak{N}_{n-1}(Z_2)$ defined by $[T, M^n]_2 \to [T \mid V^{n-1}, V^{n-1}]_2$ is a well-defined function for $N > n$ independent of N. The resulting $\Delta : \mathfrak{N}_*(Z_2) \to \mathfrak{N}_*(Z_2)$ is an $\mathfrak{N}$-module homomorphism of degree -1.*

Proof. Since (A, S^N) is $(N-1)$-universal for the group Z_2, for $N \geq n$ there is an equivariant map $f : (T, M^n) \to (A, S^N)$ and a commutative diagram

$$\begin{array}{ccc} M^n & \xrightarrow{\,f\,} & S^N \\ \downarrow{\scriptstyle \nu} & & \downarrow{\scriptstyle \nu} \\ M^n/T & \xrightarrow{\,\bar{f}\,} & S^N/A = P_N . \end{array}$$

By (10.1) there is a map $\bar{g} : M^n/T \to P_N$ homotopic to $\bar{f}$ and transverse regular on $P_{N-1} \subset P_N$. By the homotopy lifting property, there is an equivariant $g : M^n \to S^N$ with commutativity holding in

$$\begin{array}{ccc} M^n & \xrightarrow{\,g\,} & S^N \\ \downarrow{\scriptstyle \nu} & & \downarrow{\scriptstyle \nu} \\ M^n/T & \xrightarrow{\,\bar{g}\,} & S^N/T = P_N . \end{array}$$

Since the maps v are local diffeomorphisms, g is transverse regular on S^{N-1} if and only if $\bar{g}$ is transverse regular on P_{N-1}.

To show $\varDelta$ well defined for $N > n$, it is seen to be sufficient to show that if $[T, M^n]_2 = 0$ then $[T \mid V^{n-1}, V^{n-1}]_2 = 0$. Suppose then that $(T, M^n) = (T, \dot{B}^{n+1})$. Since $\pi_i(S^N) = 0$ for $1 \leq i \leq n$, we may extend g to an equivariant map $G: B^{n+1} \to S^N$. By (21.2), we may select a neighborhood U of $\dot{B}^{n+1}$ equivariantly diffeomorphic to $M^n \times [0, 1)$; we identify U with $M^n \times [0, 1)$. It is no restriction to suppose $G(x, t) = g(x)$ for $x \in M^n$, $0 \leq t < 1$.

There is now $\bar{G}: B^{n+1}/T \to P_N$, and $\bar{G}(x, t) = \bar{g}(x)$ for $x \in M^n/T$, $0 \leq t < 1$. It is then seen that $\bar{G}$ is transverse regular to P_{N-1} at all points of $(M^n/T) \times [0, 1/2]$ which map into P_{N-1}. By (10.1) there is then a $\bar{G}_1: B^{n+1}/T \to P_N$ transverse regular to P_{N-1} and with $\bar{G}_1 = \bar{G}$ on $(M^n/T) \times [0, 1/2]$.

Then $G_1: B^{n+1} \to S^N$ is transverse regular on S^{N-1} and $G_1 = G$ on $M^n \times [0, 1/2]$. Let $W^n = G^{-1}(S^{N-1})$. Then clearly $[T, W^n]_2 = [T \mid V^{n-1}, V^{n-1}]_2$. Hence $\varDelta$ is well-defined.

We leave it to the reader to show independence of N, and to show $\varDelta$ and an $\mathfrak{N}$-module homomorphism.

(26.2) *Let (T, M^n) be a differentiable fixed point free involution on a closed manifold. Let $W^n \subset M^n$ be a compact regular submanifold with boundary for which $W^n \cup T(W^n) = M^n$ and $W^n \cap T(W^n) = \dot{W}^n$. Then $\varDelta([T, M^n]_2) = [T, \dot{W}^n]_2$.*

Proof. First of all, select an equivariant differentiable $f: \dot{W}^n \to S^{N-1}$. Consider now the normal line bundle to $\dot{W}^n$. It is easy to see that it is trivial. Using the tubular neighborhoods of section 22, it is seen that there is a tubular neighborhood N of $\dot{W}^n$ with $N \cong \dot{W}^n \times (-1, 1)$ and with T on N given by $T(x, t) = (T(x), -t)$. Under these identifications, we may suppose $\dot{W}^n \times [0, 1) \subset W^n$ and $\dot{W}^n \times (-1, 0] \subset T(W^n)$. Denote by $S^0 \subset S^N$ the union of the north and south pole. Then $S^N \backslash S^0$ may be identified with $S^{N-1} \times (-1, 1)$ with $A(x, t) = (A(x), -t)$. Define now $G: M^n \to S^N$ so that $G: N \to S^N \backslash S^0$ is given by $G(x, t) = (g(x), t)$ and extend so that $G(W^n \backslash N) = $ north pole, $G(T(W^n) \backslash N) = $ south pole. Then G is transverse regular on S^{N-1}, and (25.2) follows from (25.1).

We now turn to a homomorphism $I_*: \mathfrak{N}_*(BO(k)) \to \mathfrak{N}_*(BO(k+1))$. This homomorphism assigns to the bordism class $[\xi]_2$ of a vector space bundle $\xi: E \to V^n$ the bordism class $[\xi']_2$ of the Whitney sum $\xi': E' \to V^n$ if ξ with a trivial line bundle. Alternatively there is a natural homotopy class of maps $I: BO(k) \to BO(k+1)$ and $I_*: \mathfrak{N}_n(BO(k)) \to \mathfrak{N}_n(BO(k+1))$ is induced by I.

(26.3) *We have $I_*: \mathfrak{N}_n(BO(k)) \cong \mathfrak{N}_n(BO(k+1))$ if $n \leq k$.*

Proof. The result follows from (8.3), using the fact that $I_*: H_j(BO(k); Z_2) \cong H_j(BO(k+1); Z_2)$ for $j \leq k$. We could go on to show that I_* is always a monomorphism. We regard (26.3) as a stability theorem, asserting that $\mathfrak{N}_n(BO(k))$ is independent of k for $k \geqq n$.

(26.4) **Theorem.** *The diagram*

$$
\begin{array}{ccc}
\mathfrak{N}_n(BO(k)) & \xrightarrow{\;J\;} & \mathfrak{N}_{n+k-1}(Z_2) \\
\downarrow{\scriptstyle I_*} & & \downarrow{\scriptstyle \Delta} \\
\mathfrak{N}_n(BO(k+1)) & \xrightarrow{\;J\;} & \mathfrak{N}_{n+k}(Z_2)
\end{array}
$$

commutes.

Proof. We first translate the above into geometric language. Let $B \to V^n$ be a differentiable $(k-1)$-sphere bundle, and let (T, B) be its bundle involution. Let $B' \to V^n$ be the Whitney join of $B \to V^n$ with a trivial 0-sphere bundle, and let (T', B') be its bundle involution. We must show that $\Delta([T', B']_2) = [T, B]_2$. We leave it to the reader to apply (26.2) to show that this is the case.

27. Dimension of fixed point sets

Here we give some of our main results concerning fixed point sets of involutions. The results are far from definitive; we hope the subject will recommend itself for further study.

(27.1) **Theorem.** *Let k be a non-negative integer. There exists an integer $\varphi(k)$ such that if (T, M^n) is a differentiable involution on a closed non-bording manifold of dimension $n > \varphi(k)$, then the dimension of some component of the fixed point set F is greater than k.*

Proof. We fix k; for $n \geqq 2k$ let $M_n = \Sigma_0^k \mathfrak{N}_j(BO(n-j))$. Let $I_* : M_n \to M_{n+1}$ be the sum of the various $I_* : \mathfrak{N}_j(BO(n-j)) \to \mathfrak{N}_j(BO(n-j+1))$. It is seen from (26.3) that $I_* : M_n \cong M_{n+1}$ for all $n \geqq 2k$.

We also define $J : M_n \to \mathfrak{N}_{n-1}(Z_2)$ to be the sum of the various $J : \mathfrak{N}_j(BO(n-j)) \to \mathfrak{N}_{n-1}(Z_2)$ defined in section 25. It follows from (26.4) that the diagram

$$
\begin{array}{ccccccc}
M_{2k} & \overset{I_*}{\cong} & \cdots & \overset{I_*}{\cong} & M_n & \overset{I_*}{\approx} & M_{n+1} \cdots \\
\downarrow{\scriptstyle J} & & & & \downarrow{\scriptstyle J} & & \downarrow{\scriptstyle J} \\
\mathfrak{N}_{2k-1}(Z_2) & \overset{\Delta}{\leftarrow} & \cdots & \leftarrow & \mathfrak{N}_{n-1}(Z_2) & \overset{\Delta}{\leftarrow} & \mathfrak{N}_n(Z_2) \cdots
\end{array}
$$

is commutative.

We define a sequence of subgroups $K_{2k}, \ldots, K_n, \ldots$ of M_{2k} via the above diagram. Namely

$$
K_n = \mathrm{Ker}(J I_*^{n-2k} : M_{2k} \to \mathfrak{N}_{n-1}(Z_2)) \, .
$$

Commutativity shows that $K_n \supset K_{n+1}$. Since M_{2k} is finite, there is an n_0 with $K_n = K_{n_0}$ for all $n \geqq n_0$.

We shall now show that we may take $\varphi(k) = n_0 - 1$. Suppose that (T, V^n) is an involution on a closed manifold with $[V^n]_2 \neq 0$, $n \geq n_0$ and $\dim F \leq k$. There are the normal bundles $\xi_m : E_m \to F^m$, $m \leq k$, and $[\xi_m]_2 \in \mathfrak{N}_m(BO(n-m))$, $m \leq k$. We have $\alpha = \Sigma_{m \leq k}[\xi_m]_2 \in M_n$, and by (24.1) we have $J(\alpha) = 0$. There exists $\beta \in M_{2k}$ with $I_*^{n-2k}(\beta) = \alpha$; then $\beta \in K_n$. Since $n \geq n_0$, then $\beta \in K_{n+1}$. On the other hand, the proof of (24.2) states that $JI_*(\alpha) = [A, S^0]_2[V^n]_2 \neq 0$. Then $JI_*^{n-2k+1}(\beta) \neq 0$ so that $\beta \notin K_{n+1}$. We have a contradiction and we may take $\varphi(k) = n_0 - 1$.

The problem of estimating $\varphi(k)$ is obviously suggested; unfortunately we have no information on this problem. It is clear that more knowledge of J is needed. We turn next to manifolds of odd Euler characteristic, first giving a bundle theory proof of the following known fact [9].

(27.2) Lemma. *Let (T, V^k) be an involution on a closed manifold with fixed point set F. Then $\chi(V^k) = \chi(F) \bmod 2$ where $\varkappa(\cdot)$ denotes the Euler characteristic.*

Proof. We may take $F^k = \emptyset$ without loss of generality. Let $\xi : B \to F$ be the normal sphere bundle to F. According to (24.1), $[T, B]_2 = \Sigma_{m < n}[T, B_m]_2 = 0$. Hence $[B/T]_2 = \Sigma[B_m/T]_2 = 0$. Now $B_m/T \to F^m$, $m < k$, is a bundle with fiber P_{k-m-1}. Then $\chi(B_m/T) = \chi(F^m) \cdot \chi(P_{k-m-1})$. Since the mod 2 Euler characteristic is a bordism invariant, we get

$$\Sigma_{m < k}\chi(F^m)\,\chi(P_{k-m-1}) = 0 \bmod 2$$
$$\Sigma_{m\,\mathrm{even}}\chi(F^m) = 0 \bmod 2$$

and hence $\chi(V^k) = \chi(F) \bmod 2$, assuming k odd.

Now take k even. By (24.2) we have $[V^k]_2 = \Sigma[B'_m/T']_2$ and $\chi(V^k) = \Sigma\chi(F^m)\,\chi(P_{k-m}) \bmod 2 = \Sigma_{m\,\mathrm{even}}\chi(F^m) = \chi(F) \bmod 2$. Thus we have (27.2).

It is convenient to assign meaning to $\mathfrak{N}_n(BO(k))$ for $k \leq 0$. Set $\mathfrak{N}_n(BO(k)) = 0$ for $k < 0$ and $\mathfrak{N}_n(BO(0)) = \mathfrak{N}_n$; this checks with the interpretation of a 0-dimensional vector space bundle as a homeomorphism $\xi : E \cong X$.

(27.3) Theorem. *Let (T, M^n) be a differentiable involution on a closed manifold of odd Euler characteristic, and let $\xi_m : E_m \to F^m$ denote the normal bundle to the union F^m of the m-dimensional components of the fixed point set F. There exists an m such that $[\xi_m]_2$ is not in the image of $I_* : \mathfrak{N}_m(BO(n-m-1)) \to \mathfrak{N}_m(BO(n-m))$.*

Proof. Suppose to the contrary that each $[\xi_m]_2$ is in the image of I_*. In particular $[F^n]_2 = 0$; it is then no loss of generality to suppose $F^n = \emptyset$. For each m let $[\xi_m]_2 = I_*([\hat{\xi}_m]_2)$ where $\hat{\xi}_m : \hat{E}_m \to \hat{F}^m$ is a differentiable $(n - m - 1)$-dimensional vector space bundle. Notice in particular that $\xi_{n-1} : E_{n-1} \to F^{n-1}$ is bordant to a trivial line bundle and hence $\Delta J([\xi_{n-1}]_2) = 0$.

Let now $(\hat{T}, \hat{B}_m)$, $m \leq n - 2$, and (T, B_m), $m \leq n - 1$, be the bundle involutions associated with $\hat{\xi}_m$ and ξ_m respectively. That is, $[\hat{T}, \hat{B}_m]_2 = J([\hat{\xi}_m]_2)$, $[T, B_m]_2 = J([\xi_m]_2)$. Then

$$\Sigma_{m \leq n-2} J([\hat{\xi}_m]_2) = \Sigma_{m \leq n-2} \Delta J I_* ([\hat{\xi}_m]_2)$$
$$= \Sigma_{m \leq n-2} \Delta J ([\xi_m]_2)$$
$$= \Sigma_{m \leq n-1} \Delta J ([\xi_m]_2)$$
$$= \Delta (\Sigma_{m \leq n-1} J ([\xi_m]_2))$$
$$= 0 \text{ by (24.1).}$$

Since $\Sigma_{m \leq n-2} [\hat{T}, \hat{B}_m]_2 = 0$ it is seen that there is a differentiable involution (T, V^{n-1}) on a closed manifold whose fixed point set is $\hat{F} = \cup \hat{F}^m$, with the normal bundle to F^m being $\hat{\xi}_m$. It is noted that $[F^m]_2 = [\hat{F}^m]_2$, so that $\chi(F^m) = \chi(\hat{F}^m) \bmod 2$. Using (27.2) and the fact that odd dimensional manifolds have Euler characteristic zero, we have $1 = \chi(M^n) = \chi(F) = \chi(\hat{F}) = \chi(V^{n-1}) = 0 \bmod 2$. We then have a contradiction, and (27.3) is established.

The stability result (26.3) yields an immediate corollary. For if $F^m = \emptyset$ for $m \geq n/2$ then each I_* is an isomorphism.

(27.4) **Corollary.** *Suppose that (T, M^{2k}) is a differentiable involution on a closed manifold of odd Euler characteristic. Then some component of the fixed point set is of dimension $\geq k$.*

Actually we can use (17.3) to obtain a more precise result. The map $I : BO(r-1) \to BO(r)$ induces $I^* : H^*(BO(r); Z_2) \to H^*(BO(r-1); Z_2)$, and the kernel of I^* is the ideal generated by the Whitney class v_r. Hence by (17.3) an element $\alpha \in \mathfrak{N}_m(BO(r))$ is in the image of $I_* : \mathfrak{N}_m(BO(r-1)) \to \mathfrak{N}_m(BO(r))$ if and only if every Whitney number of α associated with classes of the form $v_{i_1} \ldots v_{i_k} v_r$ is zero. Hence we have the following.

(27.5) *Suppose that (T, M^{2k}) is a differentiable involution on a closed manifold of odd Euler characteristic. There exists an m such that some Whitney number $\langle w_{i_1} \ldots w_{i_p} v_{j_1} \ldots v_{j_q} v_{2k-m}, \sigma_m \rangle$ of the normal bundle $\xi_m : E_m \to F^m$ is non-zero.*

In the above, the w_i are Stiefel-Whitney classes of F^m and the v_j are Whitney classes of ξ_m. We go now to some very specific applications.

(27.6) **Theorem.** *Let (T, M^n), $n > 0$, be a differentiable involution on a closed manifold, with fixed point set the disjoint union of a point and a k-sphere. Then $k = 1, 2, 4$ or 8, $n = 2k$, and M^n is bordant mod 2 to the appropriate projective plane.*

Proof. Consider first the manifolds $M^{2k} = P_2(R)$, $P_2(C)$, $P_2(Q)$ and Cayley plane. There exists an involution T on M^{2k} with fixed point set the disjoint union of a point and S^k. For the first three, let $T([z_1, z_2, z_3]) = [-z_1, z_2, z_3]$ and similarly for the Cayley plane.

Suppose now that (T, M^n), $n > 0$, has fixed point set $p \cup S^k$. There is the normal sphere bundle $\xi_k : E_k \to S^k$. Now (27.5) applies to show that ξ_k has non-zero Whitney class v_{n-k}. For by (27.2), $\chi(M^n) = \chi(p \cup S^k) = 1 \bmod 2$. Hence $n - k \geqq k$ and $n \geqq 2k$. But by (27.4), $n \leqq 2k$ and hence $n = 2k$. Since now $v_k \neq 0$, MILNOR [28] shows that $k = 1, 2, 4$, or 8.

Now let $(\hat{T}, \hat{M}^{2k})$, $n = 2k$ and $k = 1, 2, 4$, or 8, be the manifold constructed·in the first paragraph. Then T and $\hat{T}$ have the same fixed point set. Moreover $\xi_k : E_k \to S^k$ and $\hat{\xi}_k : \hat{E}_k \to S^k$ have the same Whitney class $v_k \in H^k(S^k; Z_2)$. Hence $[\xi_k]_2$ and $[\hat{\xi}_k]_2 \in \mathfrak{N}_k(BO(k))$ have the same Whitney numbers and are thus bordant. It follows from (24.2) that $[M^{2k}]_2 = [\hat{M}^{2k}]_2$. The theorem follows.

There is another case in which we can be almost as specific. Note that every M^n is the fixed point set of a non-trivial involution on a $2n$-dimensional manifold, namely of $T : M^n \times M^n \to M^n \times M^n$ where $T(x, y) = (y, x)$. We now consider $M^n = P(2r)$.

(27.7) Theorem. *Suppose that (T, M^n) is a differentiable involution on a closed manifold, with fixed point set real projective space $P(2r)$ and with $n > 2r$. Then $n = 4r$ and the Whitney class of the normal bundle* $\xi : E \to P(2r)$ *is $(1 + d)^m$ where m is odd,* $\binom{m}{2r} = 1 \bmod 2$, *and d is the non-zero element of $H^1(P(2r); Z_2)$.*

Proof. Every vector space bundle over projective space has Whitney class $(1 + d)^m$. A classical proof of this is based on the theorem of WU [46] that the Whitney classes v_{2i} determine all the Whitney classes. It also follows immediately from the Grothendieck ring of orthogonal bundles $K_0(P(s))$.

Since $\chi(P(2r)) = 1$ then $\chi(M^n) = 1 \bmod 2$ by (27.2). Then $n \leqq 4r$ by (27.4). Let now $k = n - 2r$ be the dimension of the normal bundle $\xi : E \to P(2r)$. Then k is even and $k \leqq 2r$. Let $(1 + d)^m$ be the Whitney class of ξ. Since $v_k \neq 0$ by (27.5) we have $m \geqq k$.

We show now that $m > k$. Suppose on the contrary that $m = k$. Let (T, B) be the bundle involution associated with ξ. The Stiefel-Whitney class W_1 of B/T is found by (23.4) to be

$$\begin{aligned}
W_1 &= p^*(w_1) + p^*(v_1) + \binom{m}{1} c \\
&= p^*(d) + p^*(md) + mc \\
&= p^*(d)
\end{aligned}$$

since m is even. Now $W_1^{2r} c^{k-1} = p^*(d^{2r}) c^{k-1} \neq 0$ since $d^{2r} \neq 0$. Then the involution number $\langle W_1^{2r} c^{k-1}, \sigma \rangle$ of B/T is non-zero, which contradicts the fact that $[T, B]_2 = 0$. Hence $m > k$. Clearly the same proof also shows that m is odd.

It now follows that $m > 2r$, for if $2r \geqq m > k$, then we would have $v_m \neq 0$, contradicting the fact that ξ is a k-bundle.

Finally we wish to prove that $k = 2r$ and that $\binom{2r}{m} = 1 \bmod 2$. Since $v_k \neq 0$ and $v_k = \binom{m}{k} d_k$, we see that $\binom{m}{k} = 1 \bmod 2$. Hence every term in the dyadic expansion of k occurs in the dyadic expansion of m. Since m is odd and k is even, it follows that $\binom{m}{k+1} = 1 \bmod 2$. Thus if $k < 2r$ it would follow that $v_{k+1} = \binom{m}{k+1} d^{k+1} \neq 0$, contradicting the fact that ξ is a k-dimensional vector space bundle. The theorem then follows.

We do not in fact know how many of the bundles of (27.7) can occur as normal bundle to the fixed point set. For example, must the M^n of (27.7) be bordant to $[P(2r) \times P(2r)]_2$?

28. Unrestricted bordism classes of involutions

We shall consider, by way of example, the unrestricted bordism group of involutions. We consider all differentiable involutions (T, M^n) on closed manifolds. Such an involution (T, M^n) bords if and only if there is an involution (S, B^{n+1}) on a compact manifold for which $(S, \dot{B}^{n+1})$ is equivariantly diffeomorphic to (T, M^n). From two involutions (T_1, M_1^n) and (T_2, M_2^n) a disjoint union $(T, M_1^n \cup M_2^n)$ can be formed as usual. We say that (T_1, M_1^n) and (T_2, M_2^n) are bordant if and only if the disjoint union $(T, M_1^n \cup M_2^n)$ bords in the above sense. Use of (21.2) shows that bordism is an equivalence relation; the bordism class to which (T, M^n) belongs is denoted by $\{T, M^n\}$. The collection of such bordism classes is denoted by $I_n(Z_2)$. An abelian group structure, with every element of order two, is imposed on $I_n(Z_2)$ by disjoint union. We cannot identify $I_n(Z_2)$ with the bordism group of any space, but we shall compute the group.

Let $\mathfrak{M}_n = \Sigma_0^n \mathfrak{N}_m(BO(n-m))$, where $\mathfrak{N}_n(BO(0)) = \mathfrak{N}_n$. We define $i_* : I_n(Z_2) \to \mathfrak{M}_n$ as follows. For each involution (T, M^n), let F^m denote the union of the m-dimensional components of the fixed point set and let $\xi_m : E_m \to F^m$ denote the normal bundle to F^m. Define $i_*\{T, M^n\} = \Sigma_m [\xi_m]_2 \in \mathfrak{M}_n$. By (21.2), i_* is a well-defined homomorphism. We also consider the homomorphism $J : \mathfrak{M}_n \to \mathfrak{N}_{n-1}(Z_2)$, the sum of the homomorphisms $\mathfrak{N}_m(BO(n-m)) \to \mathfrak{N}_{n-1}(Z_2)$ of section 25. By definition $J(\mathfrak{N}_n) = 0$.

(28.1) **Theorem.** *The sequence*

$$0 \longrightarrow I_n(Z_2) \xrightarrow{\ i_*\ } \mathfrak{M}_n \xrightarrow{\ J\ } \mathfrak{N}_{n-1}(Z_2) \longrightarrow 0$$

is split exact.

Proof. We first define $K : \mathfrak{N}_{n-1}(Z_2) \to \mathfrak{M}_n$ and show that $JK = $ identity. From section 23 recall that every fixed point free involution

(T, V^{n-1}) admits a unique decomposition

$$[T, V^{n-1}]_2 = \Sigma_0^{n-1}[A, S^m]_2[W^{n-m-1}]_2.$$

Now let K assign to $[T, V^{n-1}]_2$ the sum $\Sigma_0^{n-1}[\xi_m] \in \mathfrak{M}_n$, where $\xi_m : E_m \to W^{n-m-1}$ is the trivial $(m+1)$-dimensional vector space bundle over W^{n-m-1}. Clearly K is well-defined and $JK = id$. Hence J is an epimorphism.

The fact that $Im\, i_* \subset \mathrm{Ker}\, J$ is just (24.1). We leave it as an exercise to show that $Im\, i_* \supset \mathrm{Ker}\, J$.

We must show that i_* is a monomorphism. To do this we shall define $\varrho : \mathfrak{M}_n \to I_n(Z_2)$ with $\varrho i_* =$ identity.

Let $\xi : E \to V^m$ be a differentiable linear $O(n-m)$-bundle. There is the Whitney sum $\xi' : E' \to V^m$ of ξ with a trivial line bundle, with fibers $R^m \times R$. Define two bundle involutions T' and S on E' by $T'(v, t) = (-v, -t)$, $S(v, t) = (-v, t)$. We restrict the involutions to the associated $(n-m)$-sphere bundle B', noting that the two involutions commute. Of course (T', B') is the bundle involution, and S induces a fiber preserving involution $(\hat{S}, B'/T')$ on the P_{n-m}-bundle B'/T' over V^m. Note that on each fiber $\hat{S}$ leaves a point and a P_{n-m-1} fixed. Let ϱ assign to $[\xi]_2$ the bordism class $\{\hat{S}, B'/T'\}$ in $I_n(Z_2)$. There results a well defined homomorphism $\varrho : \mathfrak{N}_m(BO(n-m)) \to I_n(Z_2)$. We agree that $\varrho : \mathfrak{N}_n(BO(0)) \to I_n(Z_2)$ assigns to a closed manifold the trivial involution on that manifold. We thus obtain $\varrho : \mathfrak{M}_n \to I_n(Z_2)$. We now prove the following.

(28.2) *For $n \geqq 0$, $\varrho i_* =$ identity.*

Consider a (T, M^n) and form a new involution $(T_1, M^n \times S^1)$ by $T_1(x, z) = (Tx, -z)$, which is a fixed point free involution. Introduce also $(T_2, M^n \times S^1)$ and $(T_3, M^n \times S^1)$ by $T_2(x, z) = (x, z^{-1})$ and $T_3(x, z) = (Tx, z)$. Note that all three involutions commute. Thus T_2 and T_3 induce involutions $(T', M^n \times S^1/T_1)$ and $(S, M^n \times S^1/T_1)$ on the closed manifold $M^n \times S^1/T_1$.

Let us first describe the fixed point set of T'. Note that the fixed point set of $(T_2, M^n \times S^1)$ is $M^n \times 1 \cup M^n \times -1$. The set of coincidences of T_1 and T_2 is $F \times i \cup F \times -i$. Hence the fixed point set of $(T', M^n \times S^1/T_1)$ is the disjoint union of M^n with the fixed point set F of (T, M^n). The normal bundle to M^n in $M^n \times S^1/T_1$ is a trivial line bundle, while the normal bundle to F in $M^n \times S^1/T_1$ is the Whitney sum of the normal bundle ξ to F in M^n with a trivial line bundle.

Note next that T' restricted to the normal sphere bundle to its fixed point set in $M^n \times S^1/T_1$ reduces to the bundle involution. Moreover S and T' commute. Let $W^{n+1} \subset M^n \times S^1/T_1$ be the compact submanifold with boundary, invariant under S and T', obtained by removing the interior of a tubular neighborhood about the fixed point set of T'. Since T' acts freely on W^{n+1} we obtain an involution $(\mathcal{S}, W^{n+1}/T')$ on a compact manifold with boundary. Examination of $(\mathcal{S}, W^{n+1}/T')$ then shows $\varrho i_*\{T, M^n\} = \{T, M^n\}$. This completes (28.2) and so (28.1) follows.

CHAPTER V

Differentiable actions of $(Z_2)^k$.

Here we give a beginning to the study of differentiable actions of $(Z_2)^k$ on closed manifolds M^n. An action of $(Z_2)^k$ is equivalent to a collection of involutions $T_i : M^n \to M^n$, $i = 1, \ldots, k$, with $T_i T_j = T_j T_i$. A stationary point of the action is a point fixed under all the T_i.

We start with our usual procedure, by giving the structure of the bordism module $\mathfrak{N}_*((Z_2)^k)$ of differentiable free actions $((Z_2)^k, M^n)$. We then go on in sections 30 and 31 to give what information we have on the structure of the stationary point set of actions $((Z_2)^k, M^n)$.

29. Free Actions of $(Z_2)^k$

We consider differentiable free actions $((Z_2)^k, M^n))$ on closed manifolds; sometimes the action is denoted by (τ, M^n) where $\tau : (Z_2)^k \times M^n \to M^n$ defines the action. As in section 19, there is the module $\mathfrak{N}_*((Z_2)^k)$ of bordism classes $[(Z_2)^k, M^n]_2$ of such actions.

Recall from section 6 that there is a canonical homomorphism $\chi : \mathfrak{N}_*(X) \otimes \mathfrak{N}_*(Y) \to \mathfrak{N}_*(X \times Y)$, given by $\chi([M^n, f]_2 \otimes [V^m, g]_2) = [M^n \times V^m, f \times g]_2$. The homomorphisms

$$\mathfrak{N}_*(X) \otimes \mathfrak{N}_*(Y) \otimes \mathfrak{N}_*(Z) \to \mathfrak{N}_*(X \times Y) \otimes \mathfrak{N}_*(Z) \to \mathfrak{N}_*(X \times Y \times Z)$$

$$\mathfrak{N}_*(X) \otimes \mathfrak{N}_*(Y) \otimes \mathfrak{N}_*(Z) \to \mathfrak{N}_*(X) \otimes \mathfrak{N}_*(Y \times Z) \to \mathfrak{N}_*(X \times Y \times Z)$$

obviously coincide. If we take Y to be a single point, so that $\mathfrak{N}_*(Y) = \mathfrak{N}_*$, we see that there is induced a homomorphism $\mathfrak{N}_*(X) \otimes_{\mathfrak{N}} \mathfrak{N}_*(Y) \to \to \mathfrak{N}_*(X \times Y)$; we also denote this homomorphism by χ.

(29.1) *For CW complexes X and Y, the homomorphism $\chi : \mathfrak{N}_*(X) \otimes_{\mathfrak{N}} \otimes_{\mathfrak{N}} \mathfrak{N}^*(Y) \to \mathfrak{N}^*(X \times Y)$ is an isomorphism.*

Proof. According to (8.3), $\mathfrak{N}_*(X)$ is a free $\mathfrak{N}$-module. In fact, there is a base $\{\alpha_i\}$ for $\mathfrak{N}_*(X)$ so that $\{\mu(\alpha)_i\}$ is a base for $H_*(X; Z_2)$, where $\mu : \mathfrak{N}_n(X) \to H_n(X; Z_2)$ is as in section 6. Similarly there is a base $\{\beta_j\}$ for $\mathfrak{N}_*(Y)$ so that $\{\mu(\beta_j)\}$ is a base for $H_*(Y; Z_2)$.

Consider now the commutative diagram

$$\begin{array}{ccc} \mathfrak{N}_*(X) \otimes_{\mathfrak{N}} \mathfrak{N}_*(Y) & \xrightarrow{\chi} & \mathfrak{N}_*(X \times Y) \\ \downarrow{\mu \otimes \mu} & & \downarrow{\mu} \\ H_n(X; Z_2) \otimes H_*(Y; Z_2) & \xrightarrow{\cong} & H^*(X \times Y; Z_2). \end{array}$$

It is seen that $\{\mu\chi(\alpha_i \otimes \beta_j)\}$ constitutes a base for $H_*(X \times Y; Z_2)$. Hence, from (8.3), $\{\chi(\alpha_i \otimes \beta_j)\}$ constitutes a base for $\mathfrak{N}_*(X \times Y)$ and χ is an isomorphism.

We see now from (19.1) that for finite groups G and H we have an somorphism $\chi: \mathfrak{N}_*(G) \otimes_{\mathfrak{N}} \mathfrak{N}_*(H) \to \mathfrak{N}_*(G \times H)$. Specifically

$$\chi([G, M^n]_2 \otimes [H, V^m]_2) = [G \times H, M^n \times V^m]_2$$

where the right hand side denotes the product action. Hereafter we use χ to identify the two actions.

Recall from section 23 that $\mathfrak{N}_*(Z_2)$ is a free module, with homogeneous base $\{\gamma_i : i = 0, 1, \ldots\}$ where $\gamma_i \in \mathfrak{N}_i(Z_2)$. Then $\mathfrak{N}_*((Z_2)^k)$ is a free module with base

$$\{\gamma_{i_1} \otimes \cdots \otimes \gamma_{i_k} : i_1, \ldots, i_k = 0, 1, \ldots\}.$$

30. Actions of $(Z_2)^k$ without stationary points

We prove the following.

(30.1) **Theorem.** *If $(Z_2)^k$ acts differentiably on the closed n-manifold M^n without stationary points, then $[M^n]_2 = 0$.*

For $k = 1$ the result has already been proved (see (24.2)). Suppose the theorem true for $(Z_2)^{k-1}$. Consider a differentiable action of $(Z_2)^k$ on M^n without stationary points. Write $(Z_2)^k = Z_2 \times (Z_2)^{k-1}$. Let $F \subset M^n$ denote the fixed point set of the first Z_2. Now $(Z_2)^{k-1}$ acts on F without stationary points, and $Z_2 \times (Z_2)^{k-1}$ acts on the normal bundle N to F so that the fiber map is equivariant. Since $(Z_2)^{k-1}$ has no stationary points in F, there is no fiber of the normal bundle $q: N \to F$ carried into itself by every element of $(Z_2)^{k-1}$. The generator T of the first Z_2 acts on the normal bundle N as the antipodal involution. Consider the Whitney sum of q with a trivial line bundle over F. That is, consider $q': N \times \times R \to F$. The action of $(Z_2)^k$ can be extended to $N \times R$ as follows. For $(v, t) \in N \times R$, let $T(v, t) = (-v, -t)$ while for $g \in (Z_2)^{k-1}$, $g(v, t) = (gv, t)$. The fiber map is still equivariant, so that some $g \in (Z_2)^{k-1}$ carries a given fiber into a distinct fiber. Consider now the sphere bundle $r: B' \to F$ associated with q'. There is the action of $(Z_2)^k$ on B' with T acting as the antipodal map. There is then the action of $(Z_2)^{k-1}$ on B'/T without stationary points. Hence by the inductive assumptions $[B'/T]_2 = 0$. By (24.2) $[M^n]_2 = [B'/T]_2 = 0$. The theorem follows.

31. Actions of $Z_2 \times Z_2$ with isolated stationary points

The detailed investigation of stationary points of $(Z_2)^k$ appears to be difficult. Here we content ourselves with a single deep result, an analysis of $Z_2 \times Z_2$ acting with all stationary points isolated.

Consider all linear representations of a compact Lie group G on finite dimensional real vector spaces. V. Two such representations are equivalent if there is an equivariant linear isomorphism joining the vector spaces. We call the equivalence classes *representation classes*. A class is called *irreducible* if the representations belonging to it are irreducible and of positive degree. The *degree* of a representation class is the dimension of one of the vector spaces representing it. Denote by $R_n(G)$ the vector space over Z_2 whose generators are the representation classes of degree n. Let $R(G) = \Sigma_n R_n(G)$. We agree that there is a single representation class of degree 0, so that $R_0(G) = Z_2$. Given representations G on V_1 and on V_2, there is the representation of G on $V_1 \oplus V_2$ given by $g(v_1, v_2) = (gv_1, gv_2)$. We thus obtain a product $R_m(G) \otimes R_n(G) \to R_{m+n}(G)$. That is, $R(G)$ is a graded algebra, the *representation algebra*. It is seen that it is a polynomial algebra whose generators are the irreducible representation classes.

Suppose now that G acts differentiably on a closed n-manifold M^n and that $x \in M^n$ is a stationary point. There is then the linear representation of G on the tangent space to M^n at x. Denote this representation class by $X(x)$. For each differentiable action of G on a closed manifold having just a finite number $x_1, \ldots, x_k$ of stationary points we receive $X(x_1) + \cdots + X(x_k) \in R_n(G)$. Denote by $S_n(G) \subset R_n(G)$ the set of all such $\Sigma X(x_i)$, arising from all such actions. It is easy to see that $S_n(G)$ is a subgroup of $R_n(G)$. Moreover $S(G) = \Sigma S_n(G)$ is a subalgebra of $R(G)$. For is G acts on M_1^m with stationary points $x_1, \ldots, x_k$ and on M_2^n with stationary points $y_1, \ldots, y_2$, then using the diagonal action on $M_1^m \times M_2^n$ we have $X((x_i, y_j)) = X(x_i) \cdot X(y_j)$ and $\Sigma X(x_i, y_j) = \Sigma X(x_i) \cdot \Sigma X(y_j)$. Hence $S(G)$ is a subalgebra of $R(G)$.

Consider now $Z_2 \times Z_2$, letting T_1 and T_2 be generators. There are four irreducible representation classes Y_0, Y_1, Y_2, Y_3, of degree one, represented on the line $-\infty < s < \infty$ by

$$Y_0 : T_1(s) = s,\ T_2(s)\quad = s;$$
$$Y_1 : T_1(s) = -s,\ T_2(s) = s;$$
$$Y_2 : T_1(s) = s,\ T_2(s)\quad = -s$$
$$Y_3 : T_1(s) = -s,\ T_2(s) = -s\,.$$

Thus $R(Z_2 \times Z_2)$ is the polynomial algebra $Z_2[Y_0, Y_1, Y_2, Y_3]$.

Suppose now that $Z_2 \times Z_2$ acts differentiably on M^n and that $x \in M^n$ is an isolated stationary point. Then $X(x) = Y_1^p \cdot Y_2^q \cdot Y_3^r$. Moreover, p, q, and r have the following significance. Namely, p is the dimension of the component containing x of the fixed point set of T_2, q is the dimension of the component containing x of the fixed point set of T_1, and r is the dimension of the component containing x of the fixed point set of $T_1 \cdot T_2$.

By way of example suppose $Y_1^p Y_2^q Y_3^r \in R(Z_2 \times Z_2)$, $p + q + r = n$. This representation class is represented by a linear action of $Z_2 \times Z_2$ on R_n, with 0 the only stationary point. Compactify R^n to obtain S^n. Then we have an action of $Z_2 \times Z_2$ on S^n with precisely two stationary points 0 and ∞, and $X(0) = X(\infty) = Y_1^p Y_2^q Y_3^r$.

We next note the action of $Z_2 \times Z_2$ on the real projective plane P_2, given in homogeneous coordinates by $T_1([x,y,z]) = [-x,y,z]$, $T_2([x,y,z]) = [x, -y, z]$. The stationary points of this action are $x_1 = [1, 0, 0]$, $x_2 = [0, 1, 0]$, $x_3 = [0, 0, 1]$. The corresponding representation classes are seen to be $X(x_1) = Y_1 \cdot Y_3$, $X(x_2) = Y_2 \cdot Y_3$ and $X(x_3) = Y_1 \cdot Y_2$. For example, for x_1 this is verified by using local coordinates $[1, y, z]$ in a neighborhood of x_1. Then $T_1[1, y, z] = [1, -y, -z]$, $T_2[1, y, z] = [1, -y, z]$ and hence $X(x_1) = Y_1 \cdot Y_3$. Hence $Y_1 \cdot Y_3 + Y_1 \cdot Y_2 + Y_2 \cdot Y_3 \in S(Z_2 \times Z_2)$.

(31.1) **Theorem.** *The algebra $S(Z_2 \times Z_2)$ is the polynomial subalgebra of $R(Z_2 \times Z_2)$ generated by $Y_1 \cdot Y_2 + Y_1 \cdot Y_3 + Y_2 \cdot Y_3$.*

Proof. We have already seen that the above polynomial subalgebra is contained in $R(Z_2 \times Z_2)$. Suppose now that ΣX_i is an element of $S(Z_2 \times Z_2)$, where the X_i are distinct elements of the form $Y_1^p \cdot Y_2^q \cdot Y_3^r$. There is then a differentiable action of $Z_2 \times Z_2$ on a closed manifold, such that for each X_i there are exactly an odd number of stationary points x with $X(x) = X_i$ while for each $Y_1^p \cdot Y_2^q \cdot Y_3^r$ different from all the X_i there are exactly an even number of stationary points x with $X(x) = Y_1^p \cdot Y_2^q \cdot Y_3^r$.

Suppose that there are two stationary points x_1 and x_2 in M^n with $X(x_1) = X(x_2)$. As we have seen, there is an action of $Z_2 \times Z_2$ on S^n with 0 and ∞ as stationary points and with $X(0) = X(\infty) = X(x_1) = X(x_2)$. As in section 22, we can form a manifold from $M^n \cup S^n$ as follows. Delete small invariant open neighborhoods of x_1 and y_1, and identify the result along their spherical boundaries. Proceed similarly with x_2 and y_2. We thus get a manifold M_1^n and a differentiable action of $Z_2 \times Z_2$ on M_1^n; moreover M_1^n has the same stationary points as M^n except that x_1 and X_2 have been deleted. We thus see that we may as well suppose for each i that there is exactly one fixed point x_i with $X(x_i) = X_i$.

Let $X(x_i) = Y_1^{p_i} Y_2^{q_i} Y_3^{r_i}$ where $p_i + q_i + r_i = n$. We first argue that we cannot have $X(x_i) = Y_1^n$. If $X(x_i) = Y_1^n$ then T_2 leaves every point of an invariant cellular neighborhood about x_i fixed. Since each component of $F(T_2)$ is a manifold, then $F(T_2)$ contains the component V^n of M^n containing x_i. Since T_1 has at least two fixed points on V^n, we get the contradiction $X(x_i) = X(x_j) = Y_1^n$.

Recalling that $X(x_j) = Y_1^{p_j} Y_2^{q_j} Y_3^{r_j}$, consider the set of numbers $p_1, q_1, r_1, p_2, q_2, r_2, \ldots$. Suppose to be definite that p_1 is the largest number occurring in the sequence. Of all j with $p_j = p_1$ suppose for convenience that $r_1 = \max\{r_j : p_j = p_1\}$. If $p_j = p_1$ for $j \neq 1$, then $r_j < r_1$. For otherwise $p_1 = p_j$, $r_1 = r_j$ and hence $q_1 = q_j$ so that $X(x_1) = X(x_j)$ contrary to hypothesis.

We shall now prove that n is even and $X(x_1) = Y_1^{n/2} Y^{n/2}$. That is, we prove that $p_1 = r_1 = n/2$. In order to do so, return to the action $(Z_2 \times Z_2, B)$ where B is the normal sphere bundle to $F(T_1, M^n)$. As already noted, $(Z_2 \times Z_2, B) = (Z_2 \times Z_2, C^{n+1})$ where C^{n+1} is a compact manifold and where $T_1 : C^{n+1} \to C^{n+1}$ is without fixed points. Now dimension by dimension we have $F(T_2, B) = \dot{F}(T_2, C^{n+1})$, and $F(T_2, B)/T_1 = (F(T_2, C^{n+1})/T_1)\dot{\ }$. Moreover the normal bundle to $F(T_2, B)/T_1$ in B/T_1 extends to the normal bundle to $F(T_2, B)/T_1$ in C^{n+1}/T_1. Let v_j be a Whitney class of the normal bundle to the component P_{p_j-1} of $F(T_2, B)/T_1$ in B/T_1. The element $c \in H^1(P_{p_j-1}; Z_2)$ is the characteristic class of (T_1, S^{p_j-1}). Suppose now that $p_1 > r_1$. Consider $c^{p_1-r_1-1} v_{r_1} \in H^{p_1-1}(P_{p_j-1}; Z_2)$ for all j with $p_j = p_1$. It is seen by bordism that

$$\Sigma_{p_j = p_1} \langle c^{p_1-r_1-1} v_{r_1}, \sigma(P_{p_j-1}) \rangle = 0$$

where σ denotes the fundamental class.

If $j \neq 1$ then $r_j < r_1$ and $v_{r_1} = 0$ since $v = (1 + c)^{r_j}$*. If $j = 1$ then $v_{r_1} = c^{r_1}$ and $c^{p_1-r_1-1} v_{r_1} = c^{p_1-1} \neq 0$. Hence

$$\Sigma_{p_j = p_1} \langle c^{p_1-r_1-1} v_{r_1}, \sigma(P_{p_j-1}) \rangle = 1 .$$

We have a contradiction and $r_1 = p_1$.

We show next that $q_1 = 0$. If $q_1 > 0$ there is a q_j with $q_j = q_1$ and $j \neq 1$. Then $(p_1, q_1, r_1) = (p_j, q_j, r_j)$ so that either $p_j > p_1$ or $r_j > r_1$. It is impossible that $p_j > p_1$. Since $p_1 = r_1$, it is also impossible that $r_j > r_1$. Hence $q_1 = 0$. That is, n is even and $X(x_1) = Y_1^{n/2} Y_3^{n/2}$.

We may repeat the above argument with the role of r_j and q_j interchanged, and with T_2 replaced by $T_1 T_2$. We obtain that there is a j with $X(x_j) = Y_1^{n/2} Y_2^{n/2}$. Suppose to be definite that $j = 2$.

Finally consider $\max\{q_j : r_j = n/2\}$. Let q_3 denote this maximum and consider $(p_3, q_3, n/2)$. We can use the argument above with T_2 replacing T_1 and $T_1 T_2$ replacing T_2 to show $q_3 = n/2$.

We have now that $n = 2m$ and that $X(x_1) = Y_1^m Y_3^m$, $X(x_2) = Y_1^m Y_2^m$, $X(x_3) = Y_2^m Y_3^m$. Consider also the action $(Z_2 \times Z_2, P_2)$ constructed just prior to (31.1). There is the diagonal action of $Z_2 \times Z_2$ on $(P_2)^m$. If $y_1, y_2, y_3, \ldots$ are the stationary points of this action, then $\Sigma X(y_i) = (Y_1 Y_2 + Y_1 Y_3 + Y_2 Y_3)^m$. We may as well suppose that $X(y_1) = Y_1^m Y_3^m$, $X(y_2) = Y_1^m Y_2^m$, $X(y_3) = Y_2^m Y_3^m$.

---

* We leave it to the reader to show $v = (1 + c)^{r_j}$. We also leave it to the reader to show that if $0 < k < n$ there exists an even number of i with $q_i = k$.

Delete the interiors of invariant cellular neighborhoods of x_1 and y_1 and identify along the boundary; proceed similarly for x_2 and y_2, and x_3 and y_3. Now $Z_2 \times Z_2$ acts with isolated stationary points $z_1, z_2, \ldots$ on the resulting manifold, and there are no isolated fixed points of type $Y_1^m Y_3^m$, $Y_1^m Y_2^m$, or $Y_2^m Y_3^m$. It follows from our argument that $\Sigma X(z_i) = 0$. Hence

$$\Sigma X(x_i) = \Sigma X(y_i) = (Y_1 Y_2 + Y_1 Y_3 + Y_2 Y_3)^m.$$

The theorem is proved.

(31.2) **Theorem.** *Suppose that $Z_2 \times Z_2$ acts differentiably on the closed n-manifold M^n with isolated stationary points, say $x_1, x_2, \ldots$. Either M^n bords mod 2, or else $n = 2m$ and M^n is bordant to the product $(P_2)^m$ of projective planes. If M^n bords, then $\Sigma X(x_i) = 0$; that is, there are an even number of stationary points of any type $Y_1^p Y_2^q Y_3^r$. If $[M^n]_2 = [P_2]_2^m$, then $\Sigma X(x_i) = (Y_1 Y_2 + Y_1 Y_3 + Y_2 Y_3)^m$. In particular, there are then an odd number of stationary points, and also an odd number of each of the types $Y_1^m Y_2^m$, $Y_1^m Y_3^m$, $Y_2^m Y_3^m$.*

Proof. Either $\Sigma X(x_i) = 0$ or $n = 2m$ and $\Sigma X(x_i) = (Y_1 Y_2 + Y_1 Y_3 + Y_2 Y_3)^m$ by (31.1). If $\Sigma X(x_i) = 0$, there are then an even number of stationary points of each type $Y_1^p Y_2^q Y_3^r$. We then obtain spherical actions $(Z_2 \times Z_2, S_j^n)$, each with precisely two stationary points, and such that the tubular neighborhoods of the stationary point sets of M^n and of $U_j S_j^n$ are equivariantly diffeomorphic. By the unoriented version of (22.1), $[M^n]_2 = \Sigma[S_j^n]_2 = 0$.

Suppose next that $\Sigma X(x_i) = (Y_1 Y_2 + Y_1 Y_3 + Y_2 Y_3)^m$. By adding an appropriate number of spherical actions to each of M^n and $(P_2)^m$, we get two actions whose stationary point sets have equivariantly diffeomorphic tubular neighborhoods. Since $[S^n]_2 = 0$, it follows that $[M^n]_2 = [P_2]_2^m$. The theorem follows.

The following actions of $Z_2 \times Z_2$ may be noted. For example, $Z_2 \times Z_2$ acts on the complex projective plane precisely as it did on the real projective plane. Similarly it acts on the quaternionic projective plane and on the Cayley plane. In each of these cases there are precisely three fixed points. That $[P_2(C)]_2 = [P_2]_2^2$, $[P_2(Q)]_2 = [P_2]_2^4$, now follows from (31.2). These are, of course, well-known statements.

We are in general ignorant about properties of the stationary point set of $(Z_2)^k$. We can, however, prove the following easy fact.

(31.3) **Theorem.** *If $(Z_2)^k$ acts differentiably on a closed n-manifold with $n > 0$, then there cannot be precisely one stationary point.*

Proof. Suppose that $x \in M^n$ is a stationary point. As pointed out in section 22, there is a local coordinate system around x in which $(Z_2)^k$ acts orthogonally. By representation theory, some $(Z_2)^{k-1} \subset (Z_2)^k$ has a set of stationary points of positive dimension. Let F denote the set of stationary points of $(Z_2)^{k-1}$ acting on M^n. The component C of F containing x is of positive dimension and is invariant under $(Z_2)^k$. Now $(Z_2)^k = Z_2 \times (Z_2)^{k-1}$; let T be a generator of Z_2. Then $T : C \to C$ has a fixed point x. From (25.1), T cannot have precisely one fixed point. There is a $y \in C$ with $y \neq x$ and $T(y) = y$. Then y is a stationary point, and the theorem follows.

It is easy to see by example that results of the kind proved here do not hold for actions of Z_4, that is for maps of period 4. For example, Z_4 acts on every P_{2k} with precisely one fixed point. On the solid ball $I^{2k} \subset R^{2k}$, define T of period 4 by $T(z_1, \ldots, z_k) = (iz_1, \ldots, iz_k)$ where $i = \sqrt{-1}$. Identifying antipodal points of the boundary, we get $T : P_{2k} \to$ $\to P_{2k}$ of period 4 with precisely one fixed point. We get then actions of Z_4 on both P_4 and $P_2 \times P_2$ each having precisely one stationary point. It is seen that neighborhoods of the stationary points may be deleted and the resulting manifolds identified along their spherical boundaries. There results an action of Z_4 on a closed manifold M^4 without stationary points, and with M^4 bordant to the disjoint union $P_4 \cup P_2 \times P_2$. According to (30.4), this could not happen for $Z_2 \times Z_2$.

CHAPTER VI

Differentiable involutions and bundles.

We come back to involutions to make some observations that did not fit into the framework of Chapter IV. Given a fixed point free involution (T, B) and an n-plane bundle $r : E \to B/T$, we define another n-plane bundle $\tilde{r} : \tilde{E} \to B/T$, which we call the twist of r by (T, B). In the manner of BOREL-HIRZEBRUCH, we compute its Whitney class. In section 32, we make one application showing some of the influence of the homology of the total space on the Whitney classes of normal bundles to the fixed point set. In section 33, we give some generalizations of the famed Borsuk antipode theorems.

32. The bundle involution

We consider a certain operation connecting involutions and vector space bundles; it turns out to be a form of the tensor product. One application is a proof of the Borel-Hirzebruch theorem already assumed in section 23; other applications are given in this and the following section.

Given an m-plane bundle $q : E(q) \to X$ and an n-plane bundle $r : E(r) \to X$, we assume the existence of the tensor product $q \otimes r$, an mn-plane bundle $q \otimes r : E(q \otimes r) \to X$ [20]. The fiber $(q \otimes r)^{-1}(x)$ is just the tensor product $q^{-1}(x) \otimes r^{-1}(x)$ of the fibers. Moreover there is a map ψ of the set $\{(v, v') : q(v) = r(v')\} \subset E \times E'$ into $E(q \otimes r)$ given by

$\psi(v, v') = v \otimes v'$. We also assume that if q and r are line bundles with Whitney classes $w_1(q)$ and $w_1(r)$ respectively, then $q \otimes r$ is a line bundle with $w_1(q \otimes r) = w_1(q) + w_1(r)$.

(32.1) *Suppose that $q : E(q) \to X$ is a line bundle with Whitney class $c \in H^1(X; Z_2)$ and that $r : E(r) \to X$ is an n-plane bundle. Then the total Whitney class of the n-plane bundle $q \otimes r$ is given by $w(q \otimes r) = \Sigma_{k=0}^{n}(1 + c)^k w_{n-k}(r)$.*

Proof. For $n = 1$, this is just the formula already quoted. Consider next the case in which r is a Whitney sum $r_1 \oplus \cdots \oplus r_n$ of line bundles. Then $q \otimes r = \Sigma q \otimes r_i$ and $w(q \otimes r_i) = (1 + c) + w_1(r_i)$.

Hence $w(q \otimes r) = \Pi[(1 + c) + w_1(q_i)]$

$$= \Sigma_{k=0}^{n}(1 + c)^k \Sigma_{i_1 < \ldots < i_{n-k}} w_1(r_{i_1}) \ldots w_1(r_{i_{n-k}})$$

$$= \Sigma_k (1 + c)^k w_{n-k}(r) .$$

Consider next a universal n-plane bundle $r : E(r) \to BO(n)$. There is a map $BO(1) \times \cdots \times BO(1) \to BO(n)$ as in section 10, which induces an n-plane bundle $r' : E(r') \to BO(1) \times \cdots \times BO(1)$. It is known thas r splits into line bundles, and hence the formula holds for $w(q \otimes r')$. It is also the case that $H^*(BO(n); Z_2) \to H^*(BO(1) \times \cdots \times BO(1); Z_2)$ is a monomorphism. It is then seen by naturality that the formula holds for $w(q \otimes r)$. The assertion then follows for all r by universality.

Suppose now that we are given a fixed point free involution (T, B) and an n-plane bundle $r : E(r) \to B/T$ over the orbit space. The orbit map $v : B \to B/T$ induces an n-plane bundle $r' : E(r') \to B$. Here $E(r') \subset$ $\subset B \times E(r)$ is given by $\{(b, v') : v(b) = r(v')\}$ and r' is the restriction of projection. There is the fixed point free involution $(T', E(r'))$ given by $T'(b, v) = (T(b), -v')$. The equivariant map $r' : (T', E(r')) \to (T, B)$ induces a map of orbit spaces $\tilde{r} : E(r')/T' \to B/T$. It is seen that $\tilde{r}$ is an n-plane bundle. Given the involution (T, B) and the bundle $r : E(r) \to$ $\to B/T$, we thus get the bundle $\tilde{r} : E(\tilde{r}) \to B/T$, which we call the *twist* of r by (T, B).

(32.2) *Suppose that (T, B) is a fixed point free involution and that $r : E(r) \to B/T$ is an n-plane bundle. Consider $v : B \to B/T$ as a principal $O(1)$-bundle, and let $q : E(q) \to B/T$ be the associated line bundle. The tensor product $q \otimes r : E(q \otimes r) \to B/T$ and the twist $\tilde{r} : E(\tilde{r}) \to B/T$ of r by (T, B) are equivalent n-plane bundles.*

Proof. We define φ so that commutativity holds in

$$E(\tilde{r}) \xrightarrow{\ \varphi\ } E(q \otimes r)$$

$$\tilde{r} \searrow \qquad \swarrow {\scriptstyle q \otimes r}$$

$$B/T .$$

Consider $E(r') = \{(b, v) : v(b) = r(v)\} \subset B \times E(r)$. We may identify B with the unit 0-sphere bundle in $E(q)$, so that T is given by $T(b) = -b$. Considering $B \times E(r) \subset E(q) \times E(r)$, we have $\psi : B \times E(r) \to E(q \otimes r)$ given by $\psi(b, v) = b \otimes v$. Moreover $\psi(T'(b, v)) = \psi(-b, -v) = b \otimes v = \psi(b, v)$. Hence ψ induces a map φ of $E(\tilde{r}) = E(r')/T'$ into $E(q \otimes r)$. It is seen to be a bundle map, and the assertion follows.

We now apply the preceding to prove (23.3). Let $\xi : E(\xi) \to X$ be a differentiable linear $O(n)$-bundle. There is the unit sphere bundle $B \subset E(\xi)$ and the bundle involution (T, B) as in section 23. The map $s : B/T \to X$ is a differentiable projective $(n-1)$-space bundle; we consider the tangent bundle $\eta : E(\eta) \to B/T$ along the fiber. There is the bundle $r_2 : E(r_2) \to B$, where $E(r_2)$ consists of all pairs (v, v') in $B \times E(\xi)$ with $\xi(v) = \xi(v')$ and v' perpendicular to v. We may identify r_2 with the tangent bundle along the fiber in the sphere bundle B. Moreover $T : B \to B$ maps fibers into fibers and hence induces an involution T' of $E(r_2)$, by means of the differential of T. It is seen that $T'(v, v') = (-v, -v')$. There is the bundle $E(r_2)/T' \to B/T$, and it is seen to be isomorphic to the tangent bundle η along the fiber.

Consider now the n-plane bundle $r : E(r) \to B/T$ induced from ξ by the map $s : B/T \to X$; we have $w(r) = s^* w(\xi)$. A point of $E(r)$ is determined by an antipodal pair $\{v, -v\}$ from the unit sphere of a fiber $\xi^{-1}(x)$ together with a point $v' \in \xi^{-1}(x)$. Clearly $r = r_1 \oplus r_2$ where r_1 is obtained by requiring that v' be on the line determined by v, and where r_2 has v' perpendicular to v. According to the preceding paragraph, the twist $\tilde{r}_2 : E(\tilde{r}_2) \to B/T$ of r_2 by (T, B) is isomorphic to the tangent bundle along the fiber in B/T.

Let $q : E(q) \to B/T$ be the line bundle associated with $v : B \to B/T$. Then $q \otimes r = q \otimes r_1 + q \otimes r_2$. Moreover $w(q \otimes r_1) = 1 + 2c = 1$ so that $w(q \otimes r_2) = w(q \otimes r) = \Sigma(1 + c)^k \cdot w_{n-k}(\xi)$. Thus (23.3) follows.

We shall use the following application of the twist.

(32.3) *Let (T, M^l) be a differentiable fixed point free involution, and let (T', V^{m+n}) be a differentiable involution with F^m a component of the fixed point set of T'. Let $c \in H^1(M^l/T; Z_2)$ be the fundamental class of T, and let $r : E \to F^m$ be the normal bundle to F^m in V^{m+n}. Then the normal bundle to $M^l \times F^m/T \times T' = (M^l/T) \times F^m$ in $M^l \times V^{m+n}/T \times T'$ has total Whitney class given by $\Sigma_{k=0}^n (1 + c)^k \otimes w_{n-k}(r)$.*

Proof. According to section 22, we may identify $E(r)$ with an open tubular neighborhood N of F^m. Moreover, the antipodal map T''' of $E(r)$ is identified with $T' : N \to N$. Now $M^n \times N/T \times T'$ is a tubular neighborhood of $M^n \times F^m/T \times T'$. But $M^n \times N/T \times T' \cong M^n \times E(r)/T \times T'$. It is thus seen that the normal bundle to $(M^l/T) \times F^m$ is the twist $\tilde{r}$ of $1 \otimes r$ with the involution $(T \times 1, M^l \times F^m)$. But $w(\tilde{r}) = \Sigma(1 + c)^k \otimes w_{n-k}(r)$, and the assertion follows.

There is the following application of (32.3).

(32.4) Theorem. *Suppose that T is a differentiable involution on the closed manifold V^{m+n}, and that F^m is a component of the fixed point set of T. If $H^i(V^{m+n}; Z_2) = 0$ for $n - k \leq i \leq n$, then the Whitney classes v_i of the normal bundle to F^m in V^{m+n} have $v_i = 0$ for $n - k \leq i \leq n$.*

Proof. We use here the Smith-Gysin sequence for principal Z_2-bundles [*14*]. If (T', M^l) is a fixed point free involution, there is an exact sequence

$$\cdots \to H^i(M^l/T'; Z_2) \to H^i(M^l; Z_2) \to$$

$$\to H^i(M^l/T'; Z_2) \xrightarrow{\delta} H^{i+1}(M^l/T'; Z_2) \to \cdots .$$

If $f: (T', M^l) \to (T'', M''^k)$ is equivariant, there is an induced homomorphism of the exact sequence of (T'', M''^k) into that of (T', M^l). Finally $\delta: H^i(M^l/T') \to H^{i+1}(M^l/T')$ is given by $\delta(x) = x \cdot c$, where c is the fundamental class of T'.

Select $l > n$. Consider the fixed point free involutions $(A \times T, S^l \times V^{m+n})$ and $(A \times 1, S^l \times F^m)$. The inclusion $i: S^l \times F^m \subset S^l \times V^{m+n}$ is equivariant. Let $T' = A \times T$ and $T'' = A \times 1$. Note that $S^l \times F^m/T'' = P^l \times F^m$ where P^l is projective l-space.

We shall consider the commutative diagram

$$H^{n-k-1}(S^l \times V^{m+n}/T') \xrightarrow{\delta} \cdots \xrightarrow{\delta} H^{n-1}(S^l \times V^{m+n}/T') \xrightarrow{\delta} H^n(S^l \times V^{m+n}/T')$$
$$\downarrow{i^*} \qquad\qquad\qquad\qquad \downarrow{i^*} \qquad\qquad\qquad \downarrow{i^*}$$
$$H^{n-k-1}(P^l \times F^m) \xrightarrow{\delta} \cdots \xrightarrow{\delta} H^{n-1}(P^l \times F^m) \xrightarrow{\delta} H^n(P^l \times F^m)$$

where coefficients are Z_2. Note that $\delta: H^i(P^l \times F^m) \to H^{i+1}(P^l \times F^m)$ is obtained by taking the cup-product with $c \otimes 1$.

Next consider the cohomology class $\varphi_n \in H^n(S^l \times V^{m+n}/T'; Z_2)$ which is dual, under Poincare duality, to the submanifold $P^l \times F^m$. It has been shown quite generally by THOM [*39*] that $i^*(\varphi_n)$ is the n-dimensional Whitney class v_n of the normal bundle to $P^l \times F^m$ in $S^l \times V^{m+n}/T'$. By (32.3), $i^*(\varphi_n) = c^n \otimes 1 + c^{n-1} \otimes v_1 + \cdots + 1 \otimes v_n$.

Suppose now that $H^i(V^{m+n}; Z_2) = 0$ for $n - k \leq i \leq n$. Then $H^i(S^l \times V^{m+n}; Z_2) = 0$ for $n - k \leq i \leq n$. By the Smith-Gysin sequence, $\delta: H^{i-1}(S^l \times V^{m+n}/T') \to H^i(S^l \times V^{m+n}/T')$ is an epimorphism for $n - k \leq i \leq n$. In the diagram

$$H^{n-k-1}(S^l \times V^{m+n}/T') \xrightarrow{\delta^{k+1}} H^n(S^l \times V^{m+n}/T') .$$
$$\downarrow{i^*} \qquad\qquad\qquad\qquad \downarrow{i^*}$$
$$H^{n-k-1}(P^l \times F^m) \xrightarrow{\delta^{k+1}} H^n(P^l \times F^m)$$

we then have $\varphi = \delta^{k+1}(\gamma)$. Then

$$c^n \otimes 1 + c^{n-1} \otimes v_1 + \cdots + 1 \otimes v_n = i^*(\varphi_n)$$
$$= \delta^{k+1}(i^*(\gamma))$$
$$= (c^{n-k-1} \otimes \gamma_0 + \cdots + 1 \otimes \gamma_{n-k-1})$$
$$(c^{k+1} \otimes 1)$$
$$= c^n \otimes \gamma_0 + \cdots + c^{k+1} \otimes \gamma_{n-k-1}.$$

Hence $v_i = 0$ for $n - k \leqq i \leqq n$.

33. The Borsuk antipode theorems.

In the following, M^k denotes a differentiable manifold, not necessarily closed or compact. For any map $f: S^n \to M^k$, let $A(f) \subset S^n$ be all x with $f(x) = f(-x)$. We shall prove the following.

(33.1) Theorem. *If $f: S^n \to M^k$ where $n > k$, then $\dim A(f) \geqq n - k$. If $f: S^n \to M^n$ has $f^*: H^n(M^n; Z_2) \to H^n(S^n; Z_2)$ trivial, then $A(f) \neq \emptyset$.*

The classical Borsuk-Ulam theorem, that if $f: S^n \to R^n$ then there exists $x \in S^n$ with $f(x) = f(-x)$, results immediately since $f^* = 0$. In fact the analogous result holds with R^n replaced by any noncompact connected n-manifold M^n. The above also generalizes the following well-known theorem of BORSUK: if $f: S^n \to S^n$ has $f(-x) = -f(x)$ for all x then f is of odd degree. Finally, the result conserning maps $f: S^n \to M^k$ has some overlap with theorems of BOURGIN [10] and YANG [48] concerning maps $f: S^n \to R^k$.

We now begin the proof, considering first the case in which M^k is closed and connected. There is the fixed point free involution $(T, S^n \times \times M^k \times M^k)$ given by $T(x, y, z) = (-x, z, y)$ and $S^n \times M^k \times M^k/T$ is a closed $(n + 2k)$-manifold. Projection $S^n \times M^k \times M^k \to S^n$ yields the bundle map $\eta: S^n \times M^k \times M^k/T \to P^n$, a bundle with fiber $M^k \times M^k$ and structural group Z_2 [5]. There is the invariant subset $S^n \times \Delta$ of $S^n \times M^k \times M^k$, where Δ is the diagonal of $M^k \times M^k$, and hence $S^n \times \Delta/T = P^n \times \Delta$ is a closed $(n + k)$-submanifold of $S^n \times M^k \times M^k/T$.

Recall that the normal bundle to Δ in $M^k \times M^k$ is equivalent to the tangent bundle of $\Delta \cong M^k$. Hence, by (32.3), the k-dimensional Whitney class v_k of the normal bundle to $P^n \times \Delta$ in $S^n \times M^k \times M^k/T$ is given by $v_k = c^k \otimes 1 + c^{k-1} \otimes w_1 + \cdots + 1 \otimes w_k$, where c is the non-zero element of $H^1(P^n; Z_2)$ and the w_i are the Stiefel-Whitney classes of M^k.

Let $X = S^n \times M^k \times M^k$. Inclusion $i: P^n \times \Delta \subset X/T$ induces $i_*: H_{n+k}(P^n \times \Delta; Z_2) \to H_{n+k}(X/T)$. Let $\varphi_k \in H^k(X/T; Z_2)$ be the dual, under Poincare duality, of $i_*(\sigma)$ where σ is the fundamental class of $H_{n+k}(P^n \times \Delta)$. That is, φ_k is the cohomology class of X/T dual to the submanifold $P^n \times \Delta$. It has been shown by THOM [39] that

$$i^*(\varphi_k) = v_k = c^k \otimes 1 + c^{k-1} \otimes w_1 + \cdots + 1 \otimes w_k.$$

It is also seen that if N is a closed tubular neighborhood of $P^n \times \Delta$ in X/T, then φ is the image of σ under the composition

$$H_{n+k}(P^n \times \Delta) \cong H_{n+k}(N) \cong H^k(N, \dot{N}) \cong$$
$$\cong H^k(X/T, X/T \backslash N^0) \to H^k(X/T) .$$

It follows immediately that for any open set $U \supset P^n \times \Delta$, φ_k lies in the kernel of $H^k(X/T; Z_2) \to H^k(X/T \backslash U; Z_2)$.

We are now ready to consider maps $f : S^n \to M^k$. To each such map, we associate a cross-section s of η by $s((x)) = \big((x, f(x), f(-x))\big)$, where $(x) \in P^n$ corresponds to $x \in S^n$ and $((x, f(x), f(-x)) \in X/T$ corresponds to $(x, f(x), f(-x)) \in X$. If $f_t : S^n \to M^k$ is a homotopy of f, then the corresponding cross-section homotopy of s is given by $s_t((x)) = \big((x, f_t(x),$ $f_t(-x))\big)$. To each $f : S^n \to M^k$ we associate the cohomology class $s^*(\varphi_k) \in$ $\in H^k(P^n; Z_2)$. We next prove the following.

(33.2) *If $s^*(\varphi_k) \neq 0$, then* $\dim A(f) \geq n - k$.

There is the orbit map $v : S^n \to P^n$. Let $B(f) = v(A(f))$. Note that $s : P^n \to X/T$ has $s^{-1}(P^n \times \Delta) = B(f)$. Let U be a neighborhood of $P^n \times \Delta$ in X/T. Consider the diagram

$$
\begin{array}{ccc}
H^k(X/T) & \xrightarrow{\ j^*\ } & H^k(X/T \backslash U) \\
\downarrow{\scriptstyle s^*} & & \downarrow{\scriptstyle s_1^*} \\
H^k(P^n) & \xrightarrow{\ j_1^*\ } & H^k(P^n \backslash s^{-1}U) .
\end{array}
$$

Since $j^*(\varphi_k) = 0$ then $j_1^* s^*(\varphi_k) = j_1^*(c^k) = 0$. Given a neighborhood V of $B(f)$ there is a U with $s^{-1}(U) \subset V$. Hence for every neighborhood V of $B(f)$, $H^*(P^n) \to H^*(P^n \backslash V)$ kills c^k. Pass now to Alexander-Spanier cohomology; c^k is then represented by a cocycle α_k with support in V.

Suppose now that $H^{n-k}(P^n; Z_2) \to H^{n-k}(B(f); Z_2)$ is trivial. Since we are now using continuous cohomology, there is a neighborhood V of $B(f)$ with $H^{n-k}(P^n) \to H^{n-k}(\overline{V})$ trivial. Then c^{n-k} is represented by a cocycle β_{n-k} with support in $P^n \backslash \overline{V}$. Then the cup-product $c^k \cdot c^{n-k} = c^n$ is represented by $\alpha_k \cdot \beta_{n-k}$. However $\alpha_k \cdot \beta_{n-k} = 0$, since it has empty support. Hence $c^n = 0$, we have a contradiction and $H^{n-k}(P^n; Z_2) \to$ $\to H^{n-k}(B(f); Z_2)$ is nontrivial. Hence $\dim B(f) \geq n - k$ and $\dim A(f) \geq$ $\geq n - k$.

The above line of reasoning is due originally to YANG [47]. The reader will find more details in [14]. In the notation of that paper, we have shown co-ind$_{Z_2} A(f) \geq n - k$ if $s^*(\varphi_k) \neq 0$.

Now we prove that $s^*(\varphi_k) \neq 0$ if $n > k$. In any case $s^*(\varphi_k)$ depends only on the homotopy class of $f : S^n \to M^k$. Hence we may as well take f constant on the southern hemisphere; that is, $f(E_-^n) = y_0 \in M^k$. Consider $S^{n-1} \subset S^n$ as the equator; thus $f(S^{n-1}) = y_0$. We thus have the diagram

$$\begin{array}{ccc} P^{n-1} & \xrightarrow{s_1} & P^n \times \Delta \\ \downarrow{\scriptstyle i_1} & & \downarrow{\scriptstyle i} \\ P^n & \xrightarrow{s} & X/T \end{array}$$

where $s_1((x)) = ((x, y_0, y_0))$ for $x \in P^{n-1}$. Then

$$\begin{aligned} i_1^* s^*(\varphi_k) &= s_1^* i^*(\varphi_k) \\ &= s_1^*(c^k \otimes 1 + \cdots + 1 \otimes w_k) \\ &= i_1^*(c^k) \in H^k(P^{n-1}; Z_2) . \end{aligned}$$

Since $n > k$, then $s^*(\varphi_k) \neq 0$. We apply (33.2).

Consider now $f: S^n \to M^n$ with $f_*: H^n(M^n; Z_2) \to H^n(S^n; Z_2)$ trivial; the manifold M^n is for the present required to be closed and connected. We may continue to require $f(E^n_-) = y_0 \in M^n$. Now consider the equivariant map $F: S^n \to M^n \times M^n$ given by $F(x) = (f(x), f(-x))$. It is seen that F actually maps S^n into the wedge $M^n \vee M^n = M^n \times y_0 \cup \cup y_0 \times M^n$, since either x or $-x$ is always in E^n_-. The involution $(\sigma, M^n \times M^n)$ given by $\sigma(y, z) = (z, y)$ has $M^n \vee M^n$ invariant, $M^n \vee \vee M^n/\sigma = M^n$ and $F: S^n \to M^n \vee M^n$ equivariant.

(33.3) *If $\bar{F}: P^n \to M^n$ is the map between orbit spaces induced by F, then $\bar{F}^*: H^n(M^n; Z_2) \to H^n(P^n; Z_2)$ is trivial.*

We see this via the commutative diagram

$$\begin{array}{ccccccccc} & & & & H^n(M^n, y_0) & & & & \\ \downarrow{\scriptstyle f^*=0} & & \downarrow{\scriptstyle f^*} & & \downarrow{\scriptstyle f^*} & & \downarrow{\scriptstyle \bar{F}^*} & & \downarrow{\scriptstyle \bar{F}^*} \\ H^n(S^n) & \xleftarrow{\cong} & H^n(S^n, E^n_-) & \xrightarrow{\cong} & H^n(E^n_+, S^{n-1}) & \xleftarrow{\cong} & H^n(P^n, P^{n-1}) & \xrightarrow{\cong} & H^n(P^n) . \end{array}$$

The first f^* is 0 by assumption, hence $\bar{F}^* = 0$.

(33.4) *Under the composite of*

$$P^n \times (y_0 \times y_0) \xrightarrow{i_1} S^n \times (M^n \vee M^n)/T \xrightarrow{i_2} S^n \times M^n \times M^n/T$$

we have $i_1^* i_2^*(\varphi_n) = c^n \otimes 1$.

We merely consider

$$\begin{array}{ccc} & \overset{i_1 i_2}{\nearrow} & X/T \\ P^n \times (y_0 \times y_0) & & \uparrow{\scriptstyle i} \\ & \searrow & P^n \end{array}$$

and note $i^*(\varphi_n) = c^n \otimes 1 + \cdots + 1 \otimes w_n \in H^n(P^n \times \Delta)$.

Let $Y = S^n \times (M^n \vee M^n)$, and let $s_1 : P^n \to Y/T$ be given by $s_1((x))$ $= ((x, f(x), f(-x))$. Then $s : P^n \to X/T$ is given by $i_2 s_1$. Clearly we have only to show $s_1^*(i_2^*(\varphi_n)) \neq 0$. Since s_1 is a cross-section of the fiber map $Y/T \to P^n$, then $s_1^* : H^n(Y/T; Z_2) \to H^n(P^n; Z_2)$ is an epimorphism. Select $\gamma_n \in H^n(Y/T; Z_2)$ with $s_1^*(\gamma_n) = c^n$. For example, let $\gamma_n = \eta_1^*(c^n)$ where $\eta_1 : Y/T \to P^n$ is the fiber map. Under $i_1^* : H^n(Y/T; Z_2) \to$ $\to H^n(P^n \times (y_0 \times y_0); Z_2)$, $i_1^*(\gamma_n) = c^n \otimes 1$. By (33.4) we then have $i_1^*(\gamma_n + i_2^*(\varphi_n)) = 0$ and $\gamma_n + i_2^*(\varphi_n)$ lies in the image of

$$H^n(Y/T, P^n \times (y_0 \times y_0)) \xrightarrow{\ j^*\ } H^n(Y/T) \ .$$

We now show that in the diagram

$$\begin{array}{ccc} H^n(Y/T, P^n \times (y_0 \times y_0)) & \xrightarrow{\ j^*\ } & H^n(Y/T) \\ \downarrow{\scriptstyle \beta^*} & & \downarrow{\scriptstyle s_1^*} \\ H^n(M^n, y) & \xrightarrow{\ \overline{F}^*\ } & H^n(P^n) \end{array}$$

that $s_1^* j^* = 0$. The map $\beta : Y/T = S^n \times (M^n \vee M^n)/T \to M^n \vee M^n/\sigma = M^n$ is induced by projection $S^n \times (M^n \vee M^n) \to M^n \vee M^n$. It is seen that $Y/T \backslash P^n \times (y_0 \times y_0) = S^n \times M^n \backslash S^n \times y_0$ and that β is the projection $S^n \times (M^n, y_0) \to (M^n, y_0)$. But projection induces an isomorphism $H^n(M^n, y_0) \cong H^n(S^n \times (M^n, y_0))$, and hence β^* is an isomorphism. Since $\overline{F}^* = 0$ by (33.3), it follows that $s_1^* j^* = 0$.

Now $\gamma + i_2^*(\varphi_n)$ lies in the image of j^*, thus $s_1^*(\gamma_n + i_2^*(\varphi_n)) = 0$. Hence $s^*(\varphi_n) = s_1^* i_2^*(\varphi_n) = s_1^*(\gamma_n) = c^n \otimes 1 \neq 0$. The theorem now follows for M^k closed.

The extension of the theorem to a compact manifold M^k with boundary is immediate by doubling M^k. An open manifold may be regarded as the increasing union of compact manifolds with boundary, and since S^n is compact we thus get (33.1) for open manifolds.

(33.5) Corollary. *Suppose f is a map of S^n into the non-compact connected differentiable manifold M^n. There exists $x \in S^n$ with $f(-x)$ $= f(x)$.*

This follows immediately from (33.1), since $H^n(M^n; Z_2) = 0$.

We now consider briefly the question of what generalizations we may make for the involution (A, S^n). Suppose that (T, S'^n) is a fixed point free involution on a closed manifold which is a homotopy n-sphere. There exists equivariant maps $\varphi : (T, S'^n) \to (A, S^n)$ and $\varphi' : (A, S^n) \to (T, S'^n)$ [14]. If $f : S'^n \to M^k$ let $A_T(f)$ be the set of $x \in S^n$ with $f(x)$ $= f(Tx)$. It is seen that φ' maps $A(f\varphi')$ into $A_T(f)$ equivariantly. Moreover, φ' is of odd degree. We have actually shown in the proof of (33.1) that co-ind$_{Z_2} A(f\varphi') \geq n - k$. Since $A(f\varphi')$ is mapped equivariantly into $A_T(f)$, we get co-ind$_{Z_2} A(f) \geq n - k$. Hence we get (33.1) for the pair (T, S'^n).

(33.6) **Corollary.** *Any pair of fixed point free involutions T_1 and T_2 on S^n have a co-incidence.*

Proof. Consider the quotient map $\nu : S^n \to S^n/T_2$. It may be seen that $\nu^* : H^n(S^n/T_2; Z_2) \to H^n(S^n; Z_2)$ is trivial. By the preceding, there is an $x \in S^n$ with $\nu(x) = \nu(T_1 x)$. This means that $T_1 x = x$ or $T_1 x = T_2 x$. Since T_1 has no fixed points then $T_1 x = T_2 x$.

The following corollary is a special case of a theorem of MILNOR [*23*].

(33.7) **Corollary.** *If G is a finite group acting freely on S^n, then every element of G of order two lies in the center of G.*

Proof. Suppose $g \in G$ is of order two. For $h \in G$, both g and $hg\,h^{-1}$ give fixed point free involutions of S^n. Hence for some x, $g(x) = hg\,h^{-1}(x)$ by (33.6). Since G acts freely $g = hg\,h^{-1}$ and g is in the center.

It would be of interest to know to what extent (33.1) generalizes. Can all differentiability hypotheses be eliminated? Can S^n be replaced by a closed manifold which is a mod 2 homology sphere? Can M^k be replaced by non-manifolds? For example, is there a map $f : S^3 \to X$ into a 2-complex with $f(-x) \neq f(x)$ for all x?

CHAPTER VII

The structure of $\Omega_*(Z_p)$, p an odd prime.

We begin now our study of maps of odd prime period. The primary problem is to compute the structure of the group $\Omega_n(Z_p)$ of bordism classes $[T, M^n]$ where T is a fixed point free orientation preserving diffeomorphism of period p on the closed oriented manifold M^n. The bordism spectral sequence of $B(Z_p)$ is trivial, and this gives the order of the reduced groups $\tilde{\Omega}_n(Z_p)$. To obtain the precise structure is harder. It is solved here by geometric methods using certain maps of period p on $P_{p-1}(C)$ with isolated fixed points. We obtain finally in (36.5) the complete additive structure of $\Omega_*(Z_p)$. We go on in section 37 to study $\Omega_*(Z_{p^k})$.

34. Preliminaries.

We denote by (T, V^n) a closed oriented manifold V^n together with an orientation preserving diffeomorphism T of period p, p an odd prime. In this section T will have no fixed points, so that (T, V^n) represents an element of $\Omega_n(Z_p)$.

A particularly important example is (T, S^{2n-1}), where T acts on S^{2n-1} in complex coordinates by $T(z_1, \ldots, z_n) = (\varrho z_1, \ldots, \varrho z_n)$ with $\varrho = \exp(2\pi i/p)$. The union $\cup\, S^{2n-1}$ of $S^1 \subset S^3 \subset \cdots$ can be given a CW topology in which each S^{2n-1} is a skeleton. Moreover T operates on the union. Let $E(Z_p) = \cup\, S^{2n-1}$; T acting on $E(Z_p)$ makes $E(Z_p)$ a universal space for Z_p. The corresponding classifying space $B(Z_p)$ is

$E(Z_p)/T = \cup\, S^{2n-1}/T$. Moreover $B(Z_p)$ is a CW complex whose $(2n-1)$-skeleton is the generalized lens space S^{2n-1}/T.

We shall assume that the homology of Z_p is given by $H_0(Z_p, Z) = Z$, $H_{2n}(Z_p, Z) = 0$ for $n > 0$ and $H_{2n+1}(Z_p, Z) = Z_p$. In the above $B(Z_p)$, the generator of $H_{2n-1}(Z_p, Z)$ is given by the fundamental class of the lens space S^{2n-1}/T. Since S^{2n-1}/T is a closed orientable manifold, it follows immediately that $\mu : \Omega_*(B(Z_p)) \to H_*(B(Z_p), Z)$ is an epimorphism. We thus get the following as a corollary from (15.1).

(34.1) *The bordism spectral sequence of $B(Z_p)$ collapses.*

There is also a reduced bordism spectral sequence associated with $\tilde{\Omega}_*(B(Z_p)) \cong \Omega_*(B(Z_p), \text{point})$. We see that the reduced bordism spectral sequence also collapses. Note that by definition (T, V^n) represents an element of $\tilde{\Omega}_n(Z_p) \cong \tilde{\Omega}_n(B(Z_p))$ if and only if $[V^n/T] = 0$ in Ω_n. By (19.4), $[V^n] = p\,[V^n/T] = 0$. Since MILNOR has shown that Ω_n has no odd torsion, then (T, V^n) represents an element of $\tilde{\Omega}_n(Z_p)$ if and only if $[V^n] = 0$.

(34.2) *The abelian group $\tilde{\Omega}_n(Z_p)$ is 0 for n even, and of order p^t for n odd where $t = \Sigma_{i \leq n} \text{ rank } \Omega_i$.*

Proof. Consider the reduced bordism spectral sequence for $B(Z_p)$. There is a filtration

$$0 \subset J_{0,\,n} \subset J_{1,\,n-1} \subset \cdots \subset J_{n,\,0} = \tilde{\Omega}_n(B(Z_p))$$

with $J_{r,\,s}/J_{r-1,\,s+1} \cong \tilde{H}_r(B(Z_p), \Omega_s)$. Hence $J_{2j,\,s} = J_{2j-1,\,s+1}$ and order $J_{2j+1,\,s}/J_{2j-1,\,s+2} = p^{\text{rank}\,\Omega_s}$. The remark follows. Note in passing that $E^2_{r,s} = 0$ if $s \not\equiv 0 \bmod 4$.

According to our previous remarks, there is a collection $\{\alpha_i\}$ of homogeneous elements of $\tilde{\Omega}_*(Z_p)$ such that $\mu : \tilde{\Omega}_*(Z_p) \to H_*(Z_p, Z)$ has $\{\mu(\alpha_i)\}$ generating $H_*(Z_p, Z)$. For example, we may take $\alpha_i = [T, S^{2i-1}]$. More generally, it can be seen that we can take for α_i any $[T', S^{2i-1}]$. The following now is obtained from the proof of (18.1).

(34.3) *Suppose that $[T, X^{2n-1}] : n = 1, 2, \ldots$ is a collection of elements of $\tilde{\Omega}_*(Z_p)$ such that $\{\mu[T, X^{2n-1}]\}$ generates $\tilde{H}_*(Z_p, Z)$. Then $\{[T, X^{2n-1}]\}$ generates the Ω-module $\tilde{\Omega}_*(Z_p)$.*

In fact it can be seen that, in the reduced bordism spectral sequence, every element of $J_{2n+1,\,s} \subset \tilde{\Omega}_{2n+s+1}(Z_p)$ can be expressed in terms of the $[T, X^{2i+1}]$, $i \leq n$.

We summarize in the following the structure of $H^*(B(Z_p), Z_p)$.

(34.4) *For every r we have $H^r(B(Z_p), Z_p) \cong Z_p$. If d_2 is a generator of $H^2(B(Z_p), Z_p)$ then d_2^r generates $H^{2r}(B(Z_p), Z_p)$. There is a generator d_1 of $H^1(B(Z_p), Z_p)$ such that the Bockstein $\delta : H^1(B(Z_p), Z_p) \to H^2(B(Z_p), Z_p)$ maps d_1 into d_2. Moreover $d_1 d_2^r$ is a generator for $H^{2r+1}(B(Z_p), Z_p)$.*

We set $d_{2r} = d_2^r$ and $d_{2r+1} = d_1 d_2^r$, having picked d_2.

For every (T, V^n), T fixed point free, the natural homomorphism $H^*(B(Z_p), Z_p) \to H^*(V^n/T, Z_p)$ maps d_k into elements which we also denote by d_k. For every (T, V^n), define mod p characteristic numbers as follows. Denoting by $p_i \in H^{4i}(V^n/T; Z_p)$ the mod p restrictions of the Pontryagin classes of the tangent bundle, the mod p *characteristic numbers* are the integers $\bmod p \langle p_{i_1} \ldots p_{i_r} d_j, \sigma_n \rangle$ where σ denotes the orientation class of V^n/T induced by the orientation of V^n and where $4_{i_1} + \cdots + 4_{i_r} + j = n$. Just as in section 17, these are invariants of the bordism class $[T, V^n]$. It is also seen that $\mu[T, V^n] \neq 0$ if and only if $\langle d_n, \sigma_n \rangle \neq 0$. For if $\varrho : V^n/T \to B(Z_p)$ induces (T, V^n) then $\langle d_n, \sigma_n \rangle = \langle \varrho^* d_n, \sigma_n \rangle = \langle d_n, \varrho^* \sigma_n \rangle$.

(34.5) *In dimensions $n < 2p - 2$, an element $[T, V^n] \in \tilde{\Omega}_n(Z_p)$ is 0 if and only if all its* mod p *characteristic numbers are 0.*

Proof. We assert first that in dimensions $n < 2p - 2$, an element $[V^n]$ is in $p\,\Omega_n$ if and only if all the Pontryagin numbers of V^n are divisible by p. We must use now the MILNOR results on the structure of Ω_* [25, 41]. Namely, Ω/T for T the torsion of Ω is a polynomial algebra with a generator X^{4k} for each dimension $4\,k$. Moreover $\langle s_k, \sigma_{4k} \rangle \neq 0 \bmod p$, $4\,k < 2p - 2$, where $\langle s_k, \sigma_{4k} \rangle$ is the linear combination of the Pontryagin numbers obtained from the symmetric function Σt_i^k. Now $[V^n] = \Sigma_{i_1 \geq \ldots \geq i_k} a_{i_1 \ldots i_k} [X^{4i_1} \ldots X^{4i_k}] \bmod T$; suppose that the Pontryagin numbers are all divisible by p. Order the above terms as $(i_1, \ldots, i_k) > (j_1, \ldots, j_l)$ if $i_1 = j_1, \ldots, i_r = j_r, i_{r+1} > j_{r+1}$. Consider the term with largest $(i_1, \ldots, i_k)$ for which $a_{i_1 \ldots i_k} \neq 0 \bmod p$. Then $\langle s_{i_1, \ldots, i_k}, \sigma_n(V^n) \rangle = a_{i_1 \ldots i_k} s_{i_1}[X^{4i_1}] \ldots s_{i_k}[X^{4i_k}] \neq 0 \bmod p$, a contradiction. Hence $a_{i_1 \ldots i_k} = 0 \bmod p$ and $[V^n] \in p\,\Omega + T$. Since Ω has only two-torsion then $T \subset \mathrm{p}\,\Omega$ and $[V^n] \in p\,\Omega$.

Fix now a generating set $\{[T, X^{2n-1}]\}$ for $\tilde{\Omega}_*(Z_p)$ as in (34.3). Consider the function $\Sigma_{4i \leq n} \Omega_{4i} \to \tilde{\Omega}_n(Z_p)$, n odd, which maps $[V^i] \in \Omega_i$ into $[T, X^{n-i}] [V^i]$.

Using the method of (17.2), using now the numbers $\langle p_{i_1} \ldots p_{i_r} d_j, \sigma_n \rangle$, we see that for $n < 2p - 2$ the kernel of this homomorphism is contained in $\Sigma_{4i \leq n} p\,\Omega_{4i}$. That is, the image is of order at least p^t where $t = \Sigma_{i \leq n} \mathrm{rank}\,\Omega_i$. Hence the homomorphism is an epimorphism by (34.2). For the order to be correct, the kernel must be precisely $\Sigma_{4i \leq n} p\,\Omega_{4i}$. It follows that every element of $\tilde{\Omega}_n(Z_p)$, $n < 2p - 2$, is of order p.

We shall see now that (34.5) becomes false in dimensions $\geq 2p - 1$. For example, $[T, S^1] [P_{p-1}(C)]$ has all its mod p characteristic numbers equal to 0, while by the next remark it is not 0 in $\tilde{\Omega}_{2p-1}(Z_p)$.

(34.6) *Suppose that (T, X^{2n+1}) has $\mu[T, X^{2n+1}] \neq 0$ in $H_{2n+1}(Z_p, Z)$. If $[V^m] \notin p\,\Omega_m$ then $[T, X^{2n+1}] [V^m] \neq 0$ in $\tilde{\Omega}_{m+2n+1}(Z_p)$.*

Proof. Consider again the reduced bordism spectral sequence for $B(Z_p)$. The edge homomorphism $J_{2n+1,0} \to E^2_{2n+1,0}$ maps $[T, X^{2n+1}]$ into a non-zero element γ_{2n+1} of $H_{2n+1}(Z_p, Z)$. Since $E^2_{2n+1,0} \otimes \Omega_m \to E^2_{2n+1,m}$ has kernel $E^2_{2n+1,0} \otimes p\,\Omega_m$, it follows that $\gamma_{2n+1} \otimes [V^m] \neq 0$ in $E^2_{2n+1,m}$ and hence $[T, X^{2n+1}][V^m] \neq 0$.

By analogy with section 26, we now define Smith homomorphisms $\Omega_n(Z_p) \to \Omega_{n-2}(Z_p)$. Define (T_j, S^{2m+1}) by $T_j(z_1, \ldots, z_{m+1}) = (\varrho^j z_1, \ldots, \varrho^j z_{m+1})$, where $\varrho = \exp(2\pi i/p)$ and $1 \leq j < p$. Regard $S^{2m-1} \subset S^{2m+1}$ as the set of all $(0, z_2, \ldots, z_{m+1})$.

We need also to recall a remark on transverse regularity. Suppose $\varphi : V^n \to V'^m$ is a differentiable map joining closed oriented manifolds, and suppose φ is transverse regular on the closed oriented submanifold W'^{m-k} of V'^m. Then $W^{n-k} = \varphi^{-1}(W'^{m-k})$ is a closed *oriented* submanifold of V^n. First one orients the normal bundle to W'^{m-k} so that the orientation of the normal bundle followed by the orientation of the tangent bundle to W'^{m-k} yields the orientation of the bundle induced on W'^{m-k} from the tangent bundle to V'^m. Next the normal bundle to W^{n-k} is oriented so that φ preserves orientation on the normal bundle. Finally the tangent bundle to W^{n-k} is oriented to that the orientation of the normal bundle followed by the orientation of the tangent bundle yields the orientation of the bundle induced on W^{n-k} from the tangent bundle to V^n.

(34.7) *Given (T, V^n) and $2m + 1 > n$, there exists an equivariant differentiable map $\varphi : (T, V^n) \to (T_j, S^{2m+1})$ which is transverse regular on S^{2m-1}. Let $W^{n-2} = \varphi^{-1}(S^{2m-1})$, let $T' = T \,|\, W^{n-2}$, and let W^{n-2} be oriented as above. The function assigning to each bordism class $[T, V^n]$ the bordism class $[T', W^{n-2}]$ is a well-defined homomorphism $\Omega_n(Z_p) \to \Omega_{n-2}(Z_p)$, which we denote by Δ_j.*

The proof is essentially that of (26.1), and is left to the reader. It is also seen that Δ_j is an Ω-module homomorphism of degree -2. We call the homomorphisms Δ_j the Smith homomorphisms.

(34.8) *For any odd prime p, $\tilde{\Omega}_*(Z_p)$ is the submodule of p-torsion of $\Omega_*(Z_p)$.*

Proof. It follows from (34.2) that $\tilde{\Omega}_n(Z_p)$ consists solely of p-torsion. On the other hand, $\Omega_*(Z_p) \cong \Omega \oplus \tilde{\Omega}_*(Z_p)$. Since Ω has no p-torsion for p odd, the remark follows.

Since Δ_j is a homomorphism, it follows immediately from (34.8) that $\Delta_j(\tilde{\Omega}_n(Z_p)) \subset \tilde{\Omega}_{n-2}(Z_p)$.

(34.9) *The homomorphism $\Delta_j : \tilde{\Omega}_*(Z_p) \to \tilde{\Omega}_*(Z_p)$ is an epimorphism.*

Proof. Select an Ω-generating set $\{[T, X^{2m-1}]\}$ for $\tilde{\Omega}_*(Z_p)$ by $(T, X^{2m-1}) = (T_j, S^{2m-1})$. It is seen that $\Delta_j[T_j, S^{2m+1}] = [T_j, S^{2m-1}]$. The remark follows.

35. The fixed point set

In this section (T, M^n) denotes a differentiable map of odd prime period p, possibly with fixed points, preserving orientation on the closed oriented manifold M^n. We show to what extent the normal bundle to the fixed point set determines the oriented bordism class of M^n.

(35.1) **Lemma.** *Let* (T_j, S^1) *be defined by* $T_j(z) = \varrho^j \cdot z$ *where* $\varrho = \exp(2\pi i/p)$. *Then*

$$[T_j \times T, S^1 \times M^n] = [T_j \times 1, S^1 \times M^n] = [T_j, S^1][M^n]$$

in $\Omega_{n+1}(Z_p)$.

Proof. We confine our proof to the case (T_1, S^1) given by $T_1(z) = \varrho z$; the general case goes in precisely the same fashion. At the beginning of section 34 we have pointed out the universal space $(T, E(Z_p))$ with (T_1, S^1) the 1-skeleton of $(T, E(Z_p))$. The element $J_{1,n}$ in the filtration of $\tilde{\Omega}_{n+1}(Z_p)$ is the image of $j_* : \tilde{\Omega}_{n+1}(S^1/T_1) \to \tilde{\Omega}_{n+1}(B(Z_p))$ where j is the inclusion $S^1/T_1 \subset B(Z_p)$. Now S^1/T_1 is a circle which we denote by S'^1. It follows from section 6 that $\tilde{\Omega}_{n+1}(S'^1) \cong \Omega_n$ under the isomorphism $[S'^1, id][W^n] \to [W^n]$. Under j_*, the element $[S'^1, id]$ goes into $[T_1, S^1] \in \Omega_1(Z_p)$. It follows from (34.5) and (34.6) that $\mathrm{Ker}\, j_* = p\, \tilde{\Omega}_{n+1}(S'^1)$.

There is also a geometric interpretation of the isomorphism $\tilde{\Omega}_{n+1}(S'^1) \cong \Omega_n$. Namely if $f : W^{n+1} \to S'^1$ is a differentiable map of a closed oriented manifold W^{n+1} with $[W^{n+1}] = 0$, choose a regular value $x_0 \in S'^1$ and send $[W^{n+1}, f]$ into $[f^{-1}(x_0)]$. Here $f^{-1}(x_0)$ is a closed n-manifold, oriented as in section 34. A typical transverse regularity argument shows this to be well-defined. The resulting homomorphism is also seen to send $[S'^1, id][W^n]$ into $[W^n]$.

Consider now the diagonal action $(T_1 \times T, S^1 \times M^n)$. The map $T_1 \times T$ is induced by the natural fiber map $f : S^1 \times M^n/Z_p \to S^1/Z_p = S'^1$. The fiber is M^n and f is regular at all points. Now by (19.4), $p[S^1 \times M^n/Z_p] = [S^1 \times M^n] = 0$; since Ω_{n+1} has no odd torsion, then $[S^1 \times M^n/Z_p] = 0$. Thus $[S^1 \times M^n/Z_p, f] \in \tilde{\Omega}_{n+1}(S'^1)$.

Under the isomorphism $\tilde{\Omega}_{n+1}(S'^1) \cong \Omega_n$, we thus see that $[S^1 \times M^n/Z_p, f]$ maps into $[M^n]$. Hence we see that $[S^1 \times M^n/Z_p, f]$ is independent of the particular $T : M^n \to M^n$, and is a function of $[M^n]$ alone. Hence we would have obtained the same element of $\tilde{\Omega}_{n+1}(S'^1)$ had we used $T = $ identity. But for $T = id$, we obtain $[S'^1, id][M^n]$. Using the homomorphism $j_* : \tilde{\Omega}_{n+1}(S'^1) \to \tilde{\Omega}_{n+1}(Z_p)$, the conclusion follows.

We continue to consider (T, M^n) where T is a differentiable map of odd prime period p, preserving orientation on the closed oriented manifold M^n. We may suppose that M^n carries a Riemannian metric in which

T is an isometry. Let F^m denote the union of the m-dimensional components of the fixed point set F of T. There is the normal $(n - m - 1)$-sphere bundle $q: B_m \to F^m$ to F^m. Moreover B_m can be identified with the boundary $\dot{N}$ of a tubular neighborhood of F^m. Thus B_m receives an induced orientation from that of N. It follows from section 22 that, under the identification $B_m \cong \dot{N}$, the differential $dT: B_m \to B_m$ corresponds to $T|\dot{N}: \dot{N} \to \dot{N}$. We denote the resulting differentiable fixed point free map of period p by (T, B_m). It follows just as in section 24 that $\Sigma_{m < n}[T, B_m] = 0$ in $\Omega_{n-1}(Z_p)$.

Let $q: E_m \to F^m$ denote the normal vector space bundle to F^m, and let $q': E'_m \to F^m$ denote the Whitney sum of a trivial 2-dimensional vector space bundle with q. Denote by $q': B'_m \to F^m$ the corresponding $(n - m + 1)$-sphere bundle. Now $E'_m = R^2 \times E_m$; define $T': E'_m \to E'_m$ by $T' = T_1 \times T$ where $T_1(z) = \varrho z$, $\varrho = \exp(2\pi i/p)$. There is then the differentiable fixed point free map (T', B'_m) of period p.

The following is the key theorem of the chapter.

(35.2) **Theorem.** *For any (T, M^n) we have*

$$\Sigma_m[T', B'_m] = [T_1, S^1][M^n]$$

in $\Omega_{n+1}(Z_p)$.

Proof. Note first the implication of the theorem. By (34.6), the right hand side determines $[M^n]$ in $\Omega_n/p\,\Omega_n$. On the other hand, the left hand side comes directly from the fixed point set, the normal bundle and the action of T on the normal bundle. The theorem is of course an analogue of (24.2).

We now proceed with the proof. Consider $I^2 = \{z: |z| \leq 1\}$ and (T_1, I^2) given by $T_1(z) = \varrho z$. Form $(T_1 \times T, I^2 \times M^n)$ and $(T_1 \times id, I^2 \times M^n)$. Then

$$(T_1 \times T, (I^2 \times M^n)^{\textstyle\cdot}) = (T_1 \times T, S^1 \times M^n)$$
$$(T_1 \times id, (I^2 \times M^n)^{\textstyle\cdot}) = (T_1 \times id, S^1 \times M^n) \ .$$

By (35.1) there is a differentiable fixed point free (τ, B^{n+2}) with $(\tau, \dot{B}^{n+2}) = (T_1 \times T, S^1 \times M^n) \cup (T_1 \times id, -S^1 \times M^n)$. We construct a differentiable (τ', V^{n+2}), with V^{n+2} a closed oriented manifold, by judicious identification of boundaries in $(T_1 \times T, I^2 \times M^n) \cup (\tau, -B^{n+2}) \cup \cup (T_1 \times id, -I^2 \times M^n)$. The fixed point set of τ' is the union of the fixed point set F of $T_1 \times T$ and the fixed point set $-M^n$ of $T_1 \times id$. Applying the rule $\Sigma[T, B_m] = 0$ to (τ', V^{n+2}) yields the theorem.

36. The structure of $\Omega_*(Z_p)$

We use the theorem of the preceding section to prove the following.

(36.1) **Theorem.** *Every element* $[T, X^{2n-1}]$ *of* $\tilde{\Omega}_{2n-1}(Z_p)$ *with* $\mu[T, X^{2n-1}] \neq 0$ *in* $H_{2n-1}(Z_p, Z)$ *has order* p^{a+1}, *where* $a(2p - 2) < < 2n - 1 < (a + 1)(2p - 2)$.

We first indicate some preliminaries. There is the map $(T, P_{p-1}(C))$ on complex projective space, given by

$$T([z_1, \ldots, z_p]) = [z_1, \varrho z_2, \ldots, \varrho^{p-1} z_p]$$

where $\varrho = \exp(2\pi i/p)$. Now T has exactly p fixed points $x_i = [0, \ldots, 0, 1_i, 0, \ldots, 0]$. In order to use (35.2), we put appropriate local coordinates about x_i. We may use as local coordinates $(z_1, \ldots, z_{p-1}) \to [z_{p-i}, \ldots, z_{p-1}, 1, z_1, \ldots, z_{p-i-1}]$ in a neighborhood of x_i. Denote by S^{2p-3} the unit sphere of this coordinate system and by $T_i : S^{2p-3} \to S^{2p-3}$ the map $T|S^{2p-3}$. It is seen that T_i is given by $T_i(z_1, \ldots, z_{p-1}) = (\varrho z_1, \ldots, \varrho^{p-1} z_{p-1})$. If we define (T', S^{2p-1}) by $T'(z_1, \ldots, z_p) = (\varrho z_1, \varrho z_2, \varrho^2 z_3, \ldots, \varrho^{p-1} z_p)$, then it follows from (35.2) that $p[T', S^{2p-1}] = [T_1, S^1][P_{p-1}(C)]$. Since $[P_{p-1}(C)] \notin p \, \Omega_{2p-2}$, it follows from (34.6) that $p[T', S^{2p-1}] \neq 0$. Now $p^2[T', S^{2p-1}] = p[T_1, S^1][P_{p-1}(C)] = 0$, thus $[T', S^{2p-1}]$ has order p^2 in $\Omega_{2p-1}(Z_p)$. By a similar line of reasoning we now prove the theorem generally.

From (34.5) we have that $[T, X^{2n-1}]$ has order p for $2n - 1 < 2p - 2$. Let (T, S^{2k-1}) be an orthogonal fixed point free map of period p on the $(2k-1)$-sphere, with $2k - 1 < 2p - 4$. Since $p[T, S^{2k-1}] = 0$ there is a (τ, V^{2k}), where V^{2k} is a closed oriented manifold and τ is an orientation preserving diffeomorphism of period p with exactly p fixed points each in an invariant S^{2k-1} with $\tau|S^{2k-1} = T$. Moreover $[V^{2k}] \notin p \, \Omega_{2k}$, for otherwise we can use (35.2) and (34.6) to show a suitable $[T', S^{2k+1}]$ has order greater than p. Since $2k + 1 < 2p - 2$ this cannot be the case by (34.5). For each $2k - 1 < 2p - 4$ select a (T, S^{2k-1}) and a corresponding (τ, V^{2k}).

Consider next the dimension $2n - 1 = a(2p - 2) - 1$. Here for (τ, V^{2n}) choose $V^{2n} = P_{p-1}(C) \times \cdots \times P_{p-1}(C)$, a factors, and let τ be the diagonal action $T \times \cdots \times T$ of the $T : P_{p-1}(C) \to P_{p-1}(C)$ already discussed. Now (τ, V^{2n}) has exactly p^a fixed points; these are all within spheres S^{2n-1} for which $\tau|S^{2n-1}$ is independent of the particular fixed point. Let (T, S^{2n-1}) denote the common orthogonal map of period p on spheres about the fixed points. We apply (35.2) to (τ, V^{2n}) and obtain a (T', S^{2n+1}) with $p^a[T', S^{2n+1}] = [T_1, S^1][V^{2n}]$. According to MIL-NOR [25, 41], $\Omega_*/p \, \Omega_*$ is a polynomial algebra over Z_p with a generator in each dimension $4k$, and for an odd prime p the class $[P_{p-1}(C)]$ may be taken as the generator of dimension $2p - 2$. In particular $[P_{p-1}(C)]^a \notin p \, \Omega_*$; thus $p^a[T', S^{2n+1}] \neq 0$ but $p^{a+1}[T', S^{2n+1}] = 0$. So far we have shown (36.1) in dimensions $< 2p - 2$ and for a certain generator in each dimension of the form $a(2p - 2) + 1$.

Consider now $2n + 1 = a(2p - 2) + k$ with $1 \leq k < 2p - 2$ and k odd. Consider $V^{2n+2} = (P_{p-1}(C))^a \times V^{k+1}$. We have already defined maps of period p on both factors; let (τ', V^{2n+2}) be the resulting diagonal map. Then τ' has p^{a+1} fixed points, and all have equivariantly diffeomorphic neighborhoods. Let (T, S^{2n+1}) denote the common sphere about the fixed points. Applying (35.2) to (τ', V^{2n+2}) we obtain a (T', S^{2n+3}) with

$$p^{a+1}[T', S^{2n+1}] = [T_1, S^1]\,[P_{p-1}(C)]^a\,[V^{k+1}] = 0 \, ,$$

where the above vanishes since $[V^{k+1}] \in p\,\Omega_*$. Thus the order of (T, S^{2n+3}) divides p^{a+1} where $a(2p-2) + 3 \leq 2n + 3 < (a+1)(2p-2)$. We have to argue eventually that the order is precisely p^{a+1} and for all $[T, X^{2n+3}]$. We use the Smith homomorphism in order to do this. We have at this stage for each n a particular (T', S^{2n-1}) with i) $[T', S^{2n-1}]$ of order p^{a+1} for $2n - 1 = a(2p-2) + 1$, ii) $[T', S^{2n-1}]$ having order dividing p^{a+1} for $a(2p-2) + 3 \leq 2n - 1 < (a+1)(2p-2)$.

We now show that every generator $[T, X^{2n-1}]$, $2n-1 = a(2p-2)+1$, has order p^{a+1}. By (34.3),

$$[T, X^{2n-1}] = b\,[T', S^{2n-1}] + [T', S^{2n-3}]\,[V^4] + \cdots .$$

Multiplying through by p^a, $p^a[T, X^{2n-1}] = bp^a[T', S^{2n-1}]$. Since $\mu[T, X^{2n-1}] \neq 0$, then $b \neq 0 \bmod p$ and $[T, X^{2n-1}]$ is of order p^{a+1}.

We have finally to show that if $a(2p-2)+3 \leq 2n-1 < (a+1)(2p-2)$ then $[T, X^{2n-1}]$ is of order p^{a+1}. Recall the operator Δ_j of section 34. It can be shown that if $[T, X^{2n-1}]$ is a generator then so is $\Delta_j[T, X^{2n-1}]$. Successive application of Δ_j carries $[T, X^{2n-1}]$ into a generator of dimension $a(2p-2)+1$, which is of order p^{a+1}. Hence order $[T, X^{2n-1}] \geq p^{a+1}$. However $[T, X^{2n-1}]$ is a linear combination of the $[T', S^{2m-1}]$, $m \leq n$. Since the $[T', S^{2m-1}]$ have order dividing p^{a+1}, then order $[T, X^{2n-1}]$ divides p^{a+1}. The theorem now follows.

(36.2) *With the notation as in (36.1) when $2n - 1 = a(2p-2) + 1$ we have $p^a[T, X^{2n-1}] = b[T_1, S^1]\,[P_{p-1}(C)]^a$ where $b \neq 0 \bmod p$.*

Proof. In the course of the proof of (36.1), it was shown that $p^a[T, X^{2n-1}] = bp^a[T', S^{2n-1}]$ where $b \neq 0 \bmod p$, and that $p^a[T', S^{2n-1}] = [T_1, S^1]\,[P_{p-1}(C)]^a$. The result follows.

Recall that MILNOR has shown that Ω/Tor is a polynomial algebra with generators $[Y^{4k}] \in \Omega_{4k}$, $k = 1, 2, \ldots$. It also follows that for an odd prime p we may take $Y^{2p-2} = P_{p-1}(C)$. We fix an Ω-generating set $\{[T, X^{2n-1}]\}$ for $\bar\Omega_*(Z_p)$ with $\mu[T, X^{2n-1}] \neq 0$ and $\Delta_j[T, X^{2n+1}] = [T, X^{2n-1}]$ for a fixed j. Let $\Gamma(p) \subset \Omega$ be the polynomial subring generated by all $[Y^{4k}]$ with $4k \neq 2p - 2$.

(36.3) **Lemma.** *Suppose*

$$\Sigma_{i+j=n}\,[T, X^{4j+1}]\,[M^{4i}] = 0$$

where each $[M^{4i}] \in \Gamma(p)$. Then $[M^{4i}] \in p^{a+1}\Omega_{4i}$ where $a(2p-2) < 4j + + 1 < (a+1)(2p-2)$.

The proof is by induction over n. That is, we assume the result for $m < n$.

We consider

$$\Sigma_{i+j=n}[T, X^{4j+1}]\,[M^{4i}] = 0 \,. \tag{i}$$

We apply Δ_i^2 to this equation and obtain

$$\Sigma_{i+j-1=n-1}[T, X^{4(j-1)+1}]\,[M^{4i}] = 0 \,. \tag{ii}$$

By the induction hypothesis, $[M^{4i}] \in p^{a+1}\Omega_{4i}$ if $a(2p-2) + 4 < 4j + 1 < (a+1)(2p-2)$, while $[M^{4i}] \in p^{a}\Omega_{4i}$ if $4j+1 = a(2p-2) + 3$. The order of $[T, X^{4j+1}]$ is p^{a+1} where $a(2p-2) < 4j+1 < (a+1) \times (2p-2)$. Thus equation (i) reads

$$\Sigma_a [T, X^{a(2p-2)+3}]\,[M^{4n-a(2p-2)-2}] = 0 \,.$$

We have

$$[M^{4n-a(2p-2)-2}] = p^{a}\,[V^{4n-a(2b-2)-2}]$$

so

$$\Sigma_a p^{a}([T, X^{a(2p-2)+3}]\,[V^{4n-a(2p-2)-2}]) = 0 \,.$$

By (36.2) we can write, with $b_a \neq 0 \bmod p$,

$$[T_1, S^1]\,\Sigma_a b_a([P_{p-1}(C)]^a)\,[V^{4n-a(2p-2)-2}]) = 0 \,.$$

This implies

$$\Sigma_a b_a([P_{p-1}(C)]^a\,[V^{4n-a(2p-2)-2}]) \in p\,\Omega_{4n+1} \,.$$

Now $[V^{4n-a(2p-2)-2}] \in \Gamma(p)$ also, thus $[V^{4n-a(2p-2)-2}] \in p\,\Omega$ and $[M^{4n-a(2p-2)-2}] \in p^{a+1}\Omega$.

We fix an integer $4n+1$. We define a_i by the rule $a_i(2p-2) < 4(n-i)+1 < (a_i+1)(2p-2)$. We consider $\Sigma_0^n \Gamma_{4i}(p)/p^{a_i+1}\Gamma_{4i}(p)$. There is a well defined homomorphism $\Sigma_0^n \Gamma_{4i}(p)/p^{a_i+1}\Gamma_{4i}(p) \to \Omega_{4n+1}(Z_p)$ which sends $[M^{4i}]$ into $[T, X^{4(n-i)+1}]\,[M^{4i}]$.

This is well defined since the order of $[T, X^{4(n-i)+1}]$ is p^{a_i+1}. The lemma (36.3) is precisely the statement that this homomorphism is a *monomorphism*. We wish to check that it is an epimorphism. The order of $\Omega_{4n+1}(Z_p)$ is p^t where $t = \Sigma_{i \leq n} \operatorname{rank} \Omega_{4i} = \Sigma_{i \leq n} s_i$ where s_i is the number of partitions of i. Let t_i be the number of partitions of i into $k_1, \ldots, k_s$ with $k_j \neq p - 1/2$ all k_j. Then $s_i = \Sigma_a t_i - a(p-1)/2$. Hence

$$\Sigma_{i \leq n} s_i = \Sigma_{i \leq n,\, a} t_{i-a(p-1)/2} = \Sigma_{j \leq n} c_j \cdot t_j \,.$$

We can compute c_j. Suppose $4j + b(2p-2) \leq 4n < 4j + (b+1) \times (2p-2)$, then we get a t_j in the sum for each $i = j + a(p-1)/2$, $a = 0, 1, \ldots, b$. Hence $c_j = b + 1$, b as above. A computation of the order of $\Sigma_0^n \Gamma_{4i}(p)/p^{a_i+1}\Gamma_{4i}(p)$ shows it is also $p^{\Sigma c_j t_j}$. Thus $\Sigma_0^n \Gamma_{4i}(p)/p^{a_i+1}\Gamma_{4i}(p)$ is isomorphic to $\Omega_{4n+1}(Z_p)$.

(36.4) *For* $n \geq 0$, $\Delta_i : \Omega_{4n+3}(Z_p) \cong \Omega_{4n+1}(Z_p)$.

We know $\Delta_i : \tilde{\Omega}_{4n+3}(Z_p) \to \tilde{\Omega}_{4n+1}(Z_p)$ is an epimorphism. The two groups have the same order, so Δ_i is an isomorphism. With (36.4) we can now give the additive structure of $\tilde{\Omega}_*(Z_p)$.

(36.5) **Theorem.** *With Ω-base $\{[T, X^{2n+1}]\}$ in $\tilde{\Omega}_*(Z_p)$ selected with* $\mu[T, X^{2n+1}] \neq 0$, $n = 0, 1, 2, \ldots$ *the group $\tilde{\Omega}_*(Z_p)$ is the direct sum of the cyclic subgroups $C_{2n+1, k_1, \ldots, k_s}$ with generators $[T, X^{2n+1}][Y_{4k_1}] \ldots$ $\ldots [Y_{4k_s}]$, one for each n and each $(k_1, \ldots, k_s)$ with $4k_j \neq 2p - 2$, all j. The order of the generator is p^{a+1}, $a(2p - 2) < 2n + 1 < (a + 1)(2p - 2)$.*

37. The bordism groups $\Omega_*(Z_{p^k})$

In this section we shall study the structure of $\tilde{\Omega}_*(Z_{p^k})$ for p an odd prime and $k \geq 1$. We are primarily concerned with computing the orders of a generating set for $\tilde{\Omega}_*(Z_{p^k})$. We shall use the fact that $\tilde{\Omega}_*(Z_p)$ has been computed, together with the transfer homomorphism studied in section 20. We shall need several results about the action of Z_{p^k} which are entirely analogous to remarks already demonstrated for Z_p. In such cases we shall only indicate the analogous proof. In this section we shall only consider free orientation preserving differentiable actions of Z_{p^k}.

Let E again be all finitely non-zero sequences of complete numbers $(z_1, z_2, \ldots)$ with $\Sigma z_1 \bar{z}_1 = 1$. Let $\lambda_k = \exp(2\pi i/p^k)$, and define (Z_{p^k}, E) by $(z_1, z_2, \ldots) \to (\lambda_k z_1, \lambda_k z_2, \ldots)$. Now $\tilde{\Omega}_*(Z_{p^k}) \cong \tilde{\Omega}_*(E/Z_{p^k})$, and we have the following.

(37.1) *For $k \geq 1$ the reduced bordism spectral sequence of $\tilde{\Omega}_*(Z_{p^k})$ collapses, $\tilde{\Omega}_{2j}(Z_{p^k}) = 0$ and the order of $\Omega_{2j+1}(Z_{p^k})$ is $(p^k)^t$ where $t = \Sigma_{4i \leq 2j+1}$ rank Ω_{4i}.*

This is entirely analogous to (34.1). Let S^{2j+1} be embedded in E as $(z_1, \ldots, z_{2(j+1)}, 0, 0, \ldots)$. Then S^{2j+1} is invariant and we let $S^{2j+1}/Z_{p^k} = L(2j + 1, p^k) \subset E/Z_{p^k}$. The image of the orientation class under $H_{2j+1}(L(2j + 1, p^k); Z) \to H_{2j+1}(Z_{p^k}, Z) = Z_{p^k}$ is the generator of $H_{2j+1}(Z_{p^k}, Z)$. There is the natural Z_p-covering map $i : E/Z_p k \to E/Z_p k + 1$ and a commutative diagram

$$L(2j + 1, p^k) \subset E/Z_{p^k}$$
$$\downarrow \qquad\qquad \downarrow i$$
$$L(2j + 1, p^{k+1}) \subset E/Z_{p^{k+1}}$$

where the first vertical map has degree p. Under $i_* : H_{2j+1}(Z_{p^k}, Z) \to \to H_{2j+1}(Z_{p^{k+1}}; Z)$ the generator of the first group goes into p times the generator of the second.

(37.2) *The map $i : E/Z_{p^k} \to E/Z_{p^{k+1}}$ induces a monomorphism $i_* : \tilde{\Omega}_*(Z_{p^k}) \to \to \tilde{\Omega}_*(Z_{p^{k+1}})$.*

This follows immediately from the fact that $i_* \otimes id : H_*(Z_{p^k}; Z) \otimes \otimes \Omega \to H_*(Z_{p^{k+1}}; Z) \otimes \Omega$ is a monomorphism together with the collapsing of the reduced bordism spectral sequences.

A Smith homomorphism $\Delta : \tilde{\Omega}_*(Z_{p^k}) \to \tilde{\Omega}_*(Z_{p^k})$ is obtained by analogy with the case $k = 1$. This Δ is an Ω-module homomorphism of degree-2. If (Z_{p^k}, S^{2j+1}) is given by $(z_1, \ldots, z_{2(j+1)}) \to (\lambda_k z_1, \ldots, \lambda_k z_{2(j+1)})$ then $\Delta([Z_{p^k}, S^{2j+1}]) = [Z_{p^k}, S^{2j-1}]$. As in (34.3) we also have the following.

(37.3) *The elements* $[Z_{p^k}, S^{2j+1}]$ *generate* $\tilde{\Omega}_*(Z_{p^k})$ *as an* Ω-module.

We only have to note that under $\mu : \tilde{\Omega}_{2j+1}(Z_{p^k}) \to H_{2j+1}(Z_{p^k}; Z)$ we have $\mu([Z_{p^k}, S^{2j+1}])$ a generator of the homology group. We see of course that $\Delta : \tilde{\Omega}_{2j+1}(Z_{p^k}) \to \tilde{\Omega}_{2j-1}(Z_{p^k})$ is an epimorphism.

(37.4) *For any* $j \geqq 0$, $\Delta : \tilde{\Omega}_{4j+3}(Z_{p^k}) \cong \tilde{\Omega}_{4j+1}(Z_{p^k})$.

We know Δ is an epimorphism. The order of $\tilde{\Omega}_{4j+3}(Z_{p^k})$ is $(p^k)^{\Sigma_{4i \leqq 4j+3} \operatorname{rank} \Omega_{4i}}$ and the order of $\tilde{\Omega}_{4j+1}(Z_{p^k})$ is $(p^k)^{\Sigma_{4i \leqq 4j+1} \operatorname{rank} \Omega_{4i}}$, but $4i \leqq 4j + 1$ if and only if $4i \leqq 4j + 3$, thus the two bordism groups have the same order so Δ is in fact an isomorphism as indicated.

We turn now to the transfer homomorphism $t : \tilde{\Omega}_*(Z_{p^{k+1}}) \to \tilde{\Omega}_*(Z_{p^k})$ defined in section 20. The transfer of a $[Z_{p^{k+1}}, M^{2j+1}]$ is obtained by taking the induced action of the subgroup $Z_{p^k} \subset Z_{p^{k+1}}$. Since $\lambda_{k+1}^p = \lambda_k$ we have $t([Z_{p^{k+1}}, S^{2j+1}]) = [Z_{p^k}, S^{2j+1}]$, and thus $t : \tilde{\Omega}_*(Z_{p^{k+1}}) \to \tilde{\Omega}_*(Z_{p^k})$ is an epimorphism. We recall that in (20.2) the composition $t i_* : \tilde{\Omega}_{2j+1}(Z_{p^k}) \to \tilde{\Omega}_{2j+1}(Z_{p^k})$ was found to be $t i_*[Z_{p^k}, M^{2j+1}] = p[Z_{p^k}, M^{2j+1}]$. Although we shall not use it, it is easy to see that $t\Delta = \Delta t$.

We compute $\tilde{\Omega}_1(Z_{p^k})$. We have $\tilde{\Omega}_1(Z_{p^k}) = J_{1,0} = E^\infty_{1,0} = E^2_{1,0} = H(Z_{p^k}, Z) = Z_{p^k}$, therefore $\mu : \tilde{\Omega}_1(Z_{p^k}) \cong H_1(Z_{p^k}; Z)$.

(37.5) *The order of* $[Z_{p^k}, S^1]$ *is* p^k *and* $i_*[Z_{p^k}, S^1] = p[Z_{p^{k+1}}, S^1]$.

The last statement is a consequence of commutativity in

$$
\begin{array}{ccc}
\tilde{\Omega}_1(Z_{p^k}) & \xrightarrow{i_*} & \tilde{\Omega}_1(Z_{p^{k+1}}) \\
\downarrow{\scriptstyle\mu} & & \downarrow{\scriptstyle\mu} \\
H_1(Z_{p^k}; Z) & \xrightarrow{i_*} & H_1(Z_{p^{k+1}}; Z) \,.
\end{array}
$$

(37.6) *If* V^n *is a closed oriented manifold then* $[Z_{p^k}, S^1][V^n] = 0$ *if and only if* $[V^n] \in p^k \Omega_n$.

Suppose we have shown this for $k < r + 1$. Now suppose $[Z_{p^r+1}, S^1] \times [V^n] = 0$. From $i_*([Z_{p^r}, S^1][V^n]) = p[Z_{p^r+1}, S^1][V^n] = 0$ we have $[Z_{p^r}, S^1][V^n] = 0$ since i_* is a monomorphism. By the inductive hypothesis $[V^n] \in p^r \Omega_n$; that is $[V^n] = p^r[M^n]$. Now $p^{r-1} i_*([Z_{p^r}, S^1][M^n]) = p^{r-1} p[Z_{p^r+1}, S^1][M^n] \equiv [Z_{p^r+1}, S^1][V^n] = 0$, thus $[Z_{p^r}, S^1][p^{r-1}M^n] = 0$. Again by induction $p^{r-1}[M^n] \in p^r \Omega_n$. That is, $p^{r-1}[M^n] = p^r[X^n]$ and $[V^n] = p^r[M^n] = p^{r+1}[X^n]$ so $[V^n] \in p^{r+1} \Omega_n$.

In view of (37.5) there is a well defined homomorphism $\Omega_{4j}/p^k\,\Omega_{4j}\to$
$\to \tilde{\Omega}_{4j+1}(Z_{p^k})$ given by $[V^{4j}] \to [Z_{p^k}, S^1]\,[V^{4j}]$. In view of (37.6) this is a
monomorphism.

(37.7) *The sequence*

$$0 \to \Omega_{4j}/p^k\,\Omega_{4j} \to \tilde{\Omega}_{4j+1}(Z_{p^k}) \xrightarrow{\Delta^2} \tilde{\Omega}_{4(j-1)+1}(Z_{p^k}) \to 0$$

is exact.

We know Δ^2 is an epimorphism. The order of its kernel is $(p^k)^{\mathrm{rank}\,\Omega_{4j}}$.
The image of $\Omega_{4j}/p^k\,\Omega_{4j}\to \tilde{\Omega}_{4j+1}(Z_{p^k})$ lies in the kernel of Δ^2 and the
order of its image is also $(p^k)^{\mathrm{rank}\,4j}$.

(37.8) *For $4j + 1 < 2p - 2$ the order of $[Z_{p^k}, S^{4j+1}]$ is p^k. If*

$$\Sigma_0^j\,[Z_{p^k}, S^{4(j-i)+1}]\,[V^{4i}] = 0, \quad 4j + 1 < 2p - 2,$$

then each $[V^{4i}] \in p^k\,\Omega_{4i}$.

The case $j = 0$ follows by (37.6). We induct on j with the aid of Δ.
Our inductive hypothesis is, with a fixed $r > 0$ and a fixed s, $4s + 1 <$
$< 2p - 2$,

   a) the result is true for $k < r + 1$ and all $4j + 1 < 2p - 2$
   b) the result is true for $k = r + 1$ and all $4j + 1 < 4s + 1$.

We must demonstrate the result for the pair $(r + 1, 4s + 1)$.

We consider $i_*\,[Z_{p^r}, S^{4s+1}] - p\,[Z_{p^{r+1}}, S^{4s+1}]$. Since $t i_*([Z_{p^r}, S^{4s+1}])$
$= p\,[Z_{p^r}, S^{4s+1s}]$ and $t\,[Z_{p^{r+1}}, S^{4s+1}] = [Z_{p^r}, S^{4s+1}]$ we have

$$t(i_*\,[Z_{p^r}, S^{4s+1}] - p\,[Z_{p^{r+1}}, S^{4s+1}]) = 0 . \tag{i}$$

Since $\mu(i_*\,[Z_{p^r}, S^{4s+1}]) = \mu(p\,[Z_{p^{r+1}}, S^{4s+1}]) = p$ times the generator of
$H_{4s+1}(Z_{p^{r+1}}, Z)$ we have

$$\mu(i_*\,[Z_{p^r}, S^{4s+1}] - p\,[Z_{p^{r+1}}, S^{4s+1}]) = 0 \tag{ii}$$

In view of (ii) we write

$$i_*\,[Z_{p^r}, S^{4s+1}] - p\,[Z_{p^{r+1}}, S^{4s+1}]$$

$$= p^{r+1}[Z_{p^{r+1}}, S^{4s+1}]\,[V^0] + \Sigma_1^s\,[Z_{p^{r+1}}, S^{4(s-i)+1}]\,[V^{4i}] \tag{iii}$$

Now $\Delta^2(p^{r+1}[Z_{p^{r+1}}, S^{4s+1}]) = p^{r+1}[Z_{p^{r+1}}, S^{4(s-1)+1}]$. By part b) of the
inductive hypothesis, $p^{r+1}[Z_{p^{r+1}}, S^{4(s-1)+1}] = 0$, thus $p^{r+1}[Z_{p^{r+1}}, S^{4s+1}] \times$
$\times [V^0]$ is in the kernel of Δ^2, so it may be replaced by $[Z_{p^{r+1}}, S^1]\,[X^{4s}]$.
We may simply write (iii) as

$$i_*\,[Z_{p^r}, S^{4s+1}] - p\,[Z_{p^{r+1}}, S^{4s+1}] = \Sigma_1^s\,[Z_{p^{r+1}}, S^{4(s-i)+1}]\,[V^{4i}] . \tag{iv}$$

We apply the transfer homomorphism to both sides of (iv) and
$\Sigma_1^s\,[Z_{p^r}, S^{4(s-i)+1}]\,[V^{4i}] = 0$. From part a) of the inductive hypothesis
we have $[V^{4i}] \in p^r\,\Omega_{4i}$ for $1 \leq i \leq s$. We write (iv) now as

$$i_*\,[Z_{p^r}, S^{4s+1}] - p\,[Z_{p^{r+1}}, S^{4s+1}] = p^r(\Sigma_1^s\,[Z_{p^{r+1}}, S^{4(s-i)+1}]\,[M^{4i}]) \tag{v}$$

We multiply (v) through by p, apply part b) of the inductive hypothesis and get

$$p\, i_*\,[Z_{p^r},\, S^{4s+1}] = p^2\,[Z_{p^{r+1}},\, S^{4s+1}]\,. \tag{vi}$$

Since i_* is a monomorphism, $v > 1$, the reader may use (vi) to see the order of $[Z_{p^{r+1}},\, S^{4s+1}]$ is exactly p^{r+1}.

Suppose now $\Sigma_1^s\,[Z_{p^{r+1}},\, S^{4(s-i)+1}]\,[V^{4i}] = 0$. We apply $\varDelta^2$ to obtain $\Sigma_0^{s-1}\,[Z_{p^{r+1}},\, S^{4(s-i-1)+1}]\,[V^{4i}] = 0$. By part b) of the inductive hypothesis, $[V^{4i}] \in p^{r+1}\,\Omega_{4i}$ for $0 \leq i \leq s-1$. Now the order of $[Z_{p^{r+1}},\, S^{4(s-i)+1}]$ is p^{r+1}, so $\Sigma_0^{s-1}\,[Z_{p^{r+1}},\, S^{4(s-i)+1}]\,[p^{r+1}M^{4i}] + [Z_{p^{r+1}},\, S^1]\,[V^{4s}]$ $= [Z_{p^{r+1}},\, S^1]\,[V^{4s}] = 0$. Now $V^{4s} \in p^{r+1}\,\Omega_{4s}$ by (37.6). The inductive step is complete. We can begin since the result is true for $k = 1,\, 4s + 1 <$ $< 2p - 2$ and for $[Z_{p^{r+1}},\, S^1]$. In addition the dimensions $4s + 3 <$ $< 2p - 2$ are covered now by (37.4).

(37.9) Theorem. *The order of $[Z_{p^k},\, S^{4j+1}]$ is p^{k+a} where $a\,(2p-2) <$ $< 4j + 1 < (a+1)\,(2p-2)$.*

We fix r and $4s + 1$. We assume

a) the result is true for $k < r + 1$ and all $4j + 1$

b) the result is true for $k = r + 1$ and all $4j + 1 < 4s + 1$.

We must exhibit the order of $(Z_{p^{r+1}},\, S^{4s+1})$. We first take a special case $4s + 1 = a\,(2p-2) + 1,\ a \geq 1$. Again

$$t\,(i_*\,[Z_{p^r},\, S^{4s+1}] - p\,[Z_{p^{r+1}},\, S^{4s+1}]) = 0$$

$$\mu\,(i_*\,[Z_{p^r},\, S^{4s+1}] - p\,[Z_{p^{r+1}},\, S^{41+1}]) = 0\,.$$

Accordingly

$$i_*\,[Z_{p^r},\, S^{4s+1}] - p\,[Z_{p^{r+1}},\, S^{4s+1}]$$

$$= p^{r+1}\,[Z_{p^{r+1}},\, S^{4s+1}]\,[V^0] + \Sigma_1^s\,[Z_{p^{r+1}},\, S^{4(s-i)+1}]\,[V^{4i}] \tag{i}$$

We apply the transfer to (i) and we have

$$p^{r+1}\,[Z_{p^r},\, S^{4s+1}]\,[V^0] + \Sigma_1^s\,[Z_{p^{r+1}},\, S^{4(s-i)+1}]\,[V^{4i}] = 0\,. \tag{ii}$$

We apply the homomorphism $\varDelta$ exactly $(a-1)\,(p-1)$ times to (ii) to obtain

$$p^{r+1}\,[Z_{p^r},\, S^{2p-2+1}]\,[V^0] + \Sigma_1^{\frac{p-1}{2}}\,[Z_{p^r},\, S^{2p-2-4i+1}]\,[V^{4i}] = 0\,.$$

By part a) of our inductive hypothesis the order of $[Z_{p^r},\, S^{2p-2+1}]$ is p^{r+1}, thus

$$\Sigma_1^{\frac{p-1}{2}}\,[Z_{p^r},\, S^{2p-2-4i+1}]\,[V^{4i}] = 0\,.$$

We apply (37.8) to conclude $[V^{4i}] \in p^r \Omega_{4i}$, $1 \leq i \leq p - 1/2$. We write (i) then as

$$i_* [Z_{p^r}, S^{4s+1}] - p [Z_{p^{r+1}}, S^{4s+1}]$$

$$= p^{r+1} [Z_{p^{r+1}}, S^{4s+1}] [V^0] + p^r \left(\Sigma_1^{\frac{p-1}{2}} [Z_{p^{r+1}}, S^{4(s-i)+1}] [M^{4i}] \right) +$$

$$+ (\Sigma_{p-1/2+1}^s [Z_{p^{r+1}}, S^{4(s-i)+1}] [V^{4i}]) \,.$$

From part b) of the inductive hypothesis the order of $[Z_{p^{r+1}}, S^{4(s-i)+1}]$ is $p^{a-1+r+1} = p^{a+r}$ for $1 \leq i \leq \frac{p-1}{2}$. Again from part b) the order of $[Z_{p^{r+1}}, S^{4(s-i)+1}]$ divides $p^{a-2+r+1} = p^{a+r-1}$ for $\frac{p-1}{2} + 1 \leq i \leq s$. We multiply the last equation by p^{a+r-1} and since $r \geq 1$

$$p^{a+r-1} i_* [Z_{p^r}, S^{4s+1}] - p^{a+r} [Z_{p^{r+1}}, S^{4s+1}]$$

$$= p^{a+2r} [Z_{p^{r+1}}, S^{4s+1}] [V^0] \,. \tag{iii}$$

By part b) the order of $[Z_{p^{r+1}}, S^{4(s-1)+1}]$ is $p^{a-1+r+1} = p^{a+r}$, thus $p^{a+r} [Z_{p^{r+1}}, S^{4s+1}] [V^0]$ lies in the kernel of Δ^2. We let $p^{a+r} [Z_{p^{r+1}}, S^{4s+1}] [V^0] = [Z_{p^{r+1}}, S^1] [X^{4s}]$. Now (iii) becomes

$$p^{a+r-1} i_* [Z_{p^r}, S^{4s+1}] - p^{a+r} [Z_{p^{r+1}}, S^{4s+1}] = p^r [Z_{p^{r+1}}, S^1] [X^{4s}] \tag{iv} \,.$$

The order of $[Z_{p^r}, S^{4s+1}]$, by part a), is p^{a+r}, thus multiplying (iv) by p gives $p^{a+r+1} [Z_{p^{r+1}}, S^{4r+1}] = 0$. We recall $r \geq 1$ and we return to (iii). Now $p^{a+2r} [Z_{p^{r+1}}, S^{4s+1}] = p^{r-1} p^{a+r+1} [Z_{p^{r+1}}, S^{4s+1}] = 0$ so we have $p^{a+r-1} i_* [Z_{p^r}, S^{4s+1}] = p^{a+r} [Z_{p^{r+1}}, S^{4s+1}]$. Since i_* is a monomorphism $p^{a+r} [Z_{p^{r+1}}, S^{4s+1}] \neq 0$.

Finally we must consider $4(s + b) + 1$, $b < p - 1/2$. We go back to

$$i_* [Z_{p^r}, S^{4(s+b)+1}] - p [Z_{p^{r+1}}, S^{4(s+b)+1}]$$

$$= p^{r+1} [Z_{p^{r+1}}, S^{4(s+b)+1}] [V^0] + \Sigma_1^{s+b} [Z_{p^{r+1}}, S^{4(s+b-i)+1}] [V^{4i}] \,.$$

By the transfer

$$p^{r+1} [Z_{p^r}, S^{4(s+b)+1}] [V^0] + \Sigma_1^{s+b} [Z_{p^r}, S^{4(s+b-i)+1}] [V^{4i}] = 0 \,.$$

We apply Δ exactly $(a - 1)(p - 1) + 2b$ times noting

$$p^{r+1} [Z_{p^r}, S^{2p-2+1}] = 0 \,.$$

Again

$$\Sigma_1^{\frac{p-1}{2}} [Z_{p^r}, S^{2p-2-4i+1}] [V^{4i}] = 0 \,.$$

Now $V^{4i} \in p^r \Omega_{4i}$, $1 \leqq i \leqq p - 1/2$. We have then

$$i_*[Z_{p^r}, S^{4(s+b)+1}] - p[Z_{p^{r+1}}, S^{4(s+b)+1}]$$

$$= p^{r+1}[Z_{p^{r+1}}, S^{4(s+b)+1}][V^0] + p^r \left(\sum_1^{\frac{p-1}{2}} [Z_{p^{r+1}}, S^{4(s+b-i)+1}][M^{4i}] \right) +$$

$$+ \sum_{\frac{p-1}{2}+1}^{s+b} [Z_{p^{r+1}}, S^{4(s+b-i)+1}][V^{4i}] .$$

Since $r \geqq 1$ we multiply by p^{a+r}, apply part b) of the hypothesis and conclude

$$p^{a+r}i_*[Z_{p^r}, S^{4(s+b)+1}] - p^{a+r+1}[Z_{p^{r+1}}, S^{4(s+b)+1}]$$

$$= p^{a+2r+1}[Z_{p^{r+1}}, S^{4(s+b)+1}][V^0] . \tag{v}$$

We may by the inductive hypothesis write,

$$p^{a+2r+1}[Z_{p^{r+1}}, S^{4(s+b)+1}] = p^r (p^{a+r+1}[Z_{p^{r+1}}, S^{4(s+b)+1}])$$

$$= p^r [Z_{p^{r+1}}, S^1][X^{4(s+b)}] .$$

Since $p^{a+r}i_*[Z_{p^r}, S^{4(s+b)+1}] = 0$ now (v) reads

$$p^{a+r+1}[Z_{p^{r+1}}, S^{4(s+b)+1}] = p^r[Z_{p^{r+1}}, S^1][X^{4(s+b)}]$$

from which

$$p^{a+r+2}[Z_{p^{r+1}}, S^{4(s+b)+1}] = 0 .$$

We look back at (v). Since $r \geqq 1$

$$p^{a+2r+1}[Z_{p^{r+1}}, S^{4(s+b)+1}][V^0] = p^{r-1}p^{a+r+2}[Z_{p^{r+1}}, S^{4(s+b)+1}][V^0] = 0 .$$

Finally $p^{a+r+1}[Z_{p^{r+1}}, S^{4(s+b)+1}] = 0$. Since $\Delta^{2b}([Z_{p^{r+1}}, S^{4(s+(b+1)}] = [Z_{p^{r+1}}, S^{4s+1}]$ the order of $[Z_{p^{r+1}}, S^{4(s+b)+1}]$ is exactly p^{a+r+1}.

The proof of (37.9) is complete. Now we have the following.

(37.10) *If* $4s + 1 = a(2p - 2) + 1$ *then* $p^{a+k}[Z_{p^{k+1}}, S^{4s+1}] = p^k b_a[Z_{p^{k+1}}, S^1][P_{p-1}(C)]^a$ *where* $b_a \neq 0 \mod p$.

As we have stated (37.10) it is, in view of (36.2), true for $k = 0$. Now as we pointed in equation (iii) of (37.9)

$$p^{a+k-1}i_*[Z_{p^k}, S^{4s+1}] = p^{a+k}[Z_{p^{k+1}}, S^{4s+1}] .$$

Now $p^{a+k-1}[Z_{p^k}, S^{4s+1}] = p^{k-1}b_a[Z_{p^k}, S^1][P_{p-1}(C)]^a$ by induction. Now by (37.5)

$$i_*[Z_{p^k}, S^1] = p[Z_{p^{k+1}}, S^1]$$

thus

$$i_*(p^{k-1}b_a[Z_{p^k}, S^1][P_{p-1}(C)]^a) = p^k b_a[Z_{p^{k+1}}, S^1][P_{p-1}(C)]^a$$

$$= p^{a+k}[Z_{p^{k+1}}, S^{4s+1}] .$$

CHAPTER VIII

Fixed points of maps of odd prime period

We consider now the fixed point sets of differentiable maps $T : M^n \to M^n$ of odd prime period p. In section 38 we analyze the normal bundle ξ to the fixed point set. It turns out to be as simple as could be expected, breaking into a Whitney sum of complex vector bundles ξ_k on which T acts as multiplication by ϱ^k, $\varrho = \exp(2\pi i/p)$.

We then illustrate some connections between the properties of such $T : M^n \to M^n$ and the structure of $\Omega_*(Z_p)$ found in Chapter VII. In section 39 we describe those bordism classes in Ω which admit representatives upon which $Z_p \times Z_p$ acts differentiably, preserving orientation, and with no stationary points. In section 40 we study those $T : M^n \to M^n$ whose fixed point sets are m-manifolds, F^m, with trivial normal bundles in an appropriate sense. There follows a purely topological section giving the structure of the ideal in Ω consisting of all those $[M^n]$ whose Pontryagin numbers are all divisible by p. We then return in section 42 to the study of those T for which the normal bundle to the fixed point is trivial; here we obtain additional insight into the module structure of $\Omega_*(Z_p)$.

38. Generalities about the normal bundle

Let H be a compact Lie group which acts differentiably on a manifold M^n. There is a Riemannian metric on M^n with respect to which H acts as a group of isometries. Denote by F^m the union of the m-dimensional components of the set of stationary points. There is the normal bundle $\xi : E \to F^m$ to F^m in M^n, and ξ can be thought of as an $O(n-m)$-bundle. Moreover, H acts on E as a group of bundle maps, mapping each fibre into itself. If we fix a fibre V_x of ξ, then on V_x we have a linear representation of H and thus a (non-unique) embedding $H \subset O(n-m)$. We shall prove in this section a theorem which implies that the structural group of ξ can be reduced, on the component containing x, to the centralizer of H in $O(n-m)$. We go on to consider the case $H = Z_p$ in detail. To handle the non-abelian case we need the following mild extension of a well known Montgomery-Zippin theorem [29, p. 216].

(38.1) Lemma. *Let $r_0 : H \to G$ be a homomorphism of the compact Lie group H into the Lie group G. For each homomorphism $r : H \to G$ sufficiently close to r_0, there exists $g \in G$ with $r = g r_0 g^{-1}$.*

Proof. Consider the Lie group $H \times G$ and the graph $K(r_0) \subset H \times G$ of r_0, where $K(r_0) = \{(x, r_0(x)) : x \in H\}$. Since $K(r_0)$ is a compact subgroup of $H \times G$, the Montgomery-Zippin result asserts that there exists a neighborhood U of $K(r_0)$ such that if K' is a closed subgroup of $H \times G$ with $K' \subset U$ then $g K' g^{-1} \subset K(r_0)$ for some $g \in H \times G$. Suppose for

$r: H \to G$ that $K(r) \subset U$. Let $(h, g) \in H \times G$ be such that $(h, g) \cdot K(r) \times$ $\times (h, g)^{-1} \subset K(r_0)$. For each $x \in H$ there is $y \in H$ with $(hxh^{-1}, gr(x) g^{-1})$ $= (y, r_0(y))$ and $gr(x) g^{-1} = r_0(h) r_0(x) r_0(h^{-1})$, so (38.1) follows.

We now consider fiber bundles $\xi: E \to X$ which are co-ordinate bundles in the sense of STEENROD, and for which the base is connected, locally connected and paracompact. We suppose that the structural group G is a compact Lie group which acts effectively on the fiber F, and F is to be locally compact. We may consider G as a subgroup of the group of homeomorphisms of F onto itself.

(38.2) **Theorem.** *Let $\xi: E \to X$ be a fiber bundle with structural group G and fiber F as above. Let H be a compact Lie group which acts on E as a group of bundle maps, taking each fiber effectively onto itself. Then the structural group of ξ can be reduced to the centralizer $C(H')$ of H' in G, where $H' \subset G$ is the subgroup of homeomorphisms of F corresponding to H under some coordinate transformation $F \to F_x$.*

Proof. Let (U_i, φ_i) be a coordinate set for ξ; that is, U_i is open in X and $\varphi_i: U_i \times F \to E$ has the usual properties. For $x \in U_i$, let $\varphi_{i,x}: F_x \to F$ denote the homeomorphism $\varphi_{i,x}(f) = \varphi_i(x, f)$. For $h \in H$, we let $h_{i,x} = \varphi_{i,x}^{-1} h \varphi_{i,x}$ and $H_{i,x} = \{h_{i,x}: h \in H\}$. Since each h acts as a bundle map, then $H_{i,x} \subset G$. Let $r_{i,x}: H \to G$ denote the homomorphism $r_{i,x}(h) = h_{i,x}$.

If $x \in U_i \cap U_j$ then there is a $g \in G$ for which $r_{j,x} = g^{-1} r_{i,x} g$, for

$$h_{j,x} = \varphi_{j,x}^{-1} h \varphi_{j,x} = (\varphi_{j,x}^{-1} \varphi_{i,x}) (\varphi_{i,x}^{-1} h \varphi_{i,x}) (\varphi_{i,x}^{-1} \varphi_{j,x}) = g_{i,j}^{-1}(x) \cdot h_{i,x} \cdot g_{i,j}(x),$$

where $g_{i,j}(x) = \varphi_{i,x}^{-1} \varphi_{j,x}$.

We consider now a coordinate neighborhood (U_i, φ_i) where U_i is connected. Fix $x_0 \in U_i$, and consider the subset $V \subset U_i$ consisting of all x for which there exists a $g \in G$ with $r_{i,x} = g r_{i,x_0} g^{-1}$. It is clear that V is closed in U_i; it follows from (38.1) that it is also open in U_i. Hence for any $x \in U_i$ there exists $g \in G$ with $r_{i,x} = g r_{i,x_0} g^{-1}$. It now follows from the connectedness of X that if $x \in U_i$ and $y \in U_j$ then there exists $g \in G$ with $r_{j,y} = g r_{i,x} g^{-1}$. We fix a homomorphism $r: H \to G$ for which each $r_{i,x}$ is conjugate in G to r.

There is the space Y of all homomorphisms of H into G which have the form grg^{-1}. The space Y is naturally homeomorphic to $G/C(H')$, where $H' = \text{Image}(r)$ and $C(H')$ is the centralizer of H' in G. For a given connected coordinate neighborhood $U_i \subset X$, there is the continuous map $f: U_i \to Y$ mapping $x \in U_i$ into $r_{i,x} \in Y$. There is the map $G \to Y$ sending g into grg^{-1}. Since $Y \cong G/C(H')$, the map $G \to Y$ has a local cross-section. If $x_0 \in U_i$ there is a connected neighborhood $V_j \subset U_i$ with $x_0 \in V_j$ and a map $x \to g_x$ of V_j into G with $g_x r g_x^{-1} = r_{i,x}$ for each $x \in V_j$. We define $\theta_j: V_j \times F \to \xi^{-1}(V_x)$ by $\theta_j(x, f) = \varphi_i(x, g_x(f))$; then

$\theta_{j,x} = \varphi_{i,x} g_x$. Hence (V_j, θ_j) is an admissible coordinate set and for each $x \in V_j$ we have

$$r_{j,x}(h) = \theta_{j,x}^{-1} h \theta_{j,x} = g_x^{-1} \varphi_{i,x}^{-1} h \varphi_{i,x} g_x = g_x^{-1} r_{i,x}(h) g_x = r(h) .$$

Thus we can find coordinate sets (V_j, θ_j) covering X such that the homomorphism $r_{j,x} : H \to G$ is independent of both j and $x \in V_j$. Now if $x \in V_i \cap V_j$ then $\theta_{i,x}^{-1} h \theta_{i,x} = \theta_{j,x}^{-1} h \theta_{j,x} = (\theta_{j,x}^{-1} \theta_{i,x}) (\theta_{i,x}^{-1} h \theta_{i,x}) (\theta_{i,x}^{-1} \theta_{j,x})$ and $r(h) = g_{ji}(x) \cdot r(h) \cdot g_{ji}^{-1}(x)$ for all $h \in H$. Hence $g_{ji}(x) \in C(H')$ and the theorem is proved.

Suppose now that $\xi : E \to X$ is a fiber bundle with group G and fiber F. Given a compact Lie group H, we ask in how many ways H can act on E as a group of bundle maps on E, taking each fibre into itself? According to (38.2) we must first select an embedding of H into G. If the structural group of ξ cannot be reduced to $C(H)$, we discard that embedding. If is can be reduced to $C(H)$ we select coordinate sets (V_j, θ_j) covering X for which $\theta_{j,x}^{-1} \theta_{i,x} \in C(H)$, all $x \in V_i \cap V_j$. Then H acts on E as a group of bundle maps via $h(\theta_{i,x}(f)) = \theta_{i,x}(h(f))$; that is, so that each $\theta_{i,x}$ is equivariant. Now (38.2) states that all such actions of H can be obtained in this way. We now consider $H = Z_p$, a cyclic group of odd prime order.

(38.3) Theorem. *Suppose that $\xi : E \to X$ is an $O(m)$ bundle with fibre R^m over a connected, locally connected, paracompact base, and that $T : E \to E$ is a map of odd prime period p which carries each fiber orthogonally onto itself leaving only the zero vector fixed. There are then linear subbundles $\xi_k : E_k \to X$ of ξ, $k = 1, \ldots, (p-1)/2$ with $\xi = \xi_1 + \cdots + \xi_{(p-1)/2}$ and there exists a complex linear structure on E_k such that $T(E_k) \subset E_k$ and $T(v) = \varrho^k v$ for $v \in E_k$ where $\varrho = \exp(2\pi i/p)$.*

Proof. According to (38.2) and the subsequent discussion, we need to know the number of orthogonal actions of Z_p on R^m in which the zero vector is the only stationary point. The irreducible representations of this type are all two-dimensional. Let T be the generator of Z_p. The irreducible representations have generators given by $T_k : R^2 \to R^2$ where T_k is given in complex coordinates by $T_k(z) = \varrho^k z$, where $\varrho = \exp(2\pi i/p)$. It is no restriction to confine ourselves to $k = 1, \ldots, (p-1)/2$.

We split $T : R^m \to R^m$ into irreducible plane representations. We see that $m = 2n$, moreover $R^m = V_1 + \cdots + V_{(p-1)/2}$ where V_k is the direct sum of plane representations on which $T = T_k$. There is a complex structure on V_k such that $T(V_k) \subset V_k$ and $T(v) = \varrho^k v$ for $v \in V_k$. We assume from representation theory that the $T_1, \ldots, T_{(p-1)/2}$ give inequivalent linear representations and that V_k is the sum of all planes $R^2 \subset R^m$ with $T(R^2) \subset R^2$ and $T | R^2$ equivalent to T_k. The centralizer $C(Z_p) = C(T)$ is now easily obtained. Suppose $f \in O(m)$ is such that $fT = Tf$. If $R^2 \subset V_k$ is such that $T(R^2) \subset R^2$ and $T | R^2$ is equivalent to

T_k, then $T(fR^2) \subset fR^2$ and $T|fR^2$ is also equivalent to T_k. Then $f(R^2) \subset V_k$, so in fact we have $f(V_k) \subset V_k$. For $v \in V_k$, $f(\varrho^k v) = \varrho^k f(v)$. Setting $\varrho^k = a + bi$,

$$af(v) + bf(iv) = af(v) + bif(v)$$

and so $f(iv) = if(v)$. That is, f is complex linear. We see that $C(Z_p)$ consists of all orthogonal f with $f(V_k) \subset V_k$ and f complex linear. Thus we have $C(Z_p) = U(n_1) \times \cdots \times U(n_{(p-1)/2})$ in $O(2n)$ where $n_1 + \cdots + n_{(p-1)/2} = n$.

By (38.2) we may as well suppose the structural group of ξ is reduced to $U(n_1) \times \cdots \times U(n_{(p-1)/2})$. Letting (U_j, θ_j) be a coordinate set, we have $\theta_{j,x}^{-1} \theta_{i,x}(V_k) \subset V_k$ and $\theta_{j,x}^{-1} \theta_{i,x}$ complex linear. Then $\theta_{i,k}(V_k) = \theta_{j,k}(V_k)$. Let $E_k = \cup_x \theta_{j,x}(V_k)$. The theorem (38.3) now follows by letting $\xi_k : E_k \to X$ have fiber $\theta_{j,x}(V_k) = F_{k,x}$. We leave to the reader the following additions to (38.3).

(38.4) *The notation of (38.3) is continued. If ξ is a differentiable bundle so is each ξ_k. If $f : E \to E$ is a map which takes fibres orthogonally onto fibers, and if $Tf = fT$, then $f(E_k) \subset E_k$ and f is complex linear.*

It follows immediately from (38.3) that if $T : M^n \to M^n$ is a differentiable map of odd prime period on a closed oriented n-manifold, then the structure group of the normal bundle to the fixed point set F can be reduced on each component of F to the unitary group. In particular, the odd Whitney classes w_{2k+1} of the normal bundle all vanish [8, p. 416].

(38.5) *Suppose $T : M^n \to M^n$ is a map of odd prime period on a closed manifold for which all the odd dimensional Stiefel-Whitney classes of the tangent bundle vanish; then on each component of the fixed point set all odd Stiefel-Whitney classes also vanish.*

Proof. Let C be a component of F with total Stiefel-Whitney class $w(C)$, and normal Whitney class $v(C)$. Similarly $w(M^n)$ is the total Stiefel-Whitney class of M^n. By the sum theorem $w(C)v(C) = i^*(w(M^n))$, and $w(C) = i^*(w(M^n)) \cdot \bar{v}(C)$ where $\bar{v}$ denotes the class with $v\bar{v} = 1$. Since $v(C)$ has no terms of odd dimension, then neither does $\bar{v}(C)$, and (38.5) follows.

The above is a variant of the fact, known to SMITH, that if M^n is orientable then so is each component of F. In fact it follows from the above argument that if $w_1(M^n) = 0$ then $w_1(C) = 0$.

We now point out a generalization of the bundle involution of Chapter IV. Let $n = n_1 + \cdots + n_{(p-1)/2}$. We shall define a homomorphism,

$$J : \Omega^m\big(B(U(n_1) \times \cdots \times U(n_{(p-1)/2}))\big) \to \Omega_{m+2n-1}(Z_p) ,$$

for each odd prime p. In doing so, we replace $\Omega_m\big(B(U(n_1) \times \cdots \times U(n_{(p-1)/2}))\big)$ by the differentiable bordism group of section 8. In order that this be meaningful, we use for $B(U(n_1) \times \cdots \times U(n_{(p-1)/2}))$ a differentiable manifold which is N-classifying for $N > m$, say a product of complex Grassman manifolds. Suppose that $f: V^m \to B(U(n_1) \times \cdots \times B(U(n_{(p-1)/2}))$ represents an element of the differentiable bordism group. There is the natural action of $U(n_1) \times \cdots \times U(n_{(p-1)/2})$ on the complex linear space $C^n = C^{n_1} \times \cdots \times C^{n_{(p-1)/2}}$. Let $\xi: E \to V^m$ be the bundle with fiber C^n induced by f. There is $T: C^n \to C^n$ defined by $T(v) = \varrho^k v$ for $v \in C^{n_k}$. Then T is in the center of $U(n_1) \times \cdots \times U(n_{p-1/2})$; hence there is induced a $T: E \to E$.

We must also orient E. It is clear that E is orientable since V^m is oriented and the fibers have their natural orientation. We orient the tangent bundle to E so that the orientation of the fiber followed by the orientation of V^m yields the orientation of E.

Let now $B \subset E$ denote the bundle of unit spheres of E. Then $T: B \to B$ is a differentiable map of period p, preserving the orientation and without fixed points. Hence $[T, B] \in \Omega_{m+2n-1}(Z_p)$. We define

$$J: \Omega_m\big(B(U(n_1) \times \cdots \times U(n_{p-1/2}))\big) \to \Omega_{m+2n-1}(Z_p)$$

to send $[V^m, f]$ into $[T, B] \in \Omega_{m+2n-1}(Z_p)$. As in Chapter IV, we can consider a $U(n_1) \times \cdots \times U(n_{p-1/2})$-bundle $\xi: E \to V^m$ as generating an element $[\xi]$ of $\Omega_m\big(B(U(n_1) \times \cdots \times U(n_{p-1/2}))\big)$. It is seen that the above J is well defined and is a homomorphism.

There is also the sum homomorphism

$$J: \Sigma\, \Omega_m\big(B(U(n_1) \times \cdots \times U(n_{p-1/2}))\big) \to \Omega_{k-1}(Z_p)$$

summed over all $m + 2(n_1 + \cdots + n_{p-1/2}) = k$. Here by definition $\Omega_{k-1}\big(B(U(0))\big) = \Omega_{k-1}$, and J maps $\Omega_{k-1}\big(B(U(0))\big)$ into 0.

Suppose now that (T, M^n) consists of an orientation preserving diffeomorphism $T: M^n \to M^n$ of period p on the closed oriented manifold M^n. Each component V of the fixed point set is orientable. The normal bundle $\xi: E \to V$ is a $U(n_1) \times \cdots \times U(n_{p-1/2})$-bundle. We orient V so that the orientation of the fibre followed by that of V yields the orientation of E, where E has the orientation of a tubular neighborhood of V in M^n.

Let $V_1, \ldots, V_r$ denote the components of F. Each normal bundle $\xi^{(j)}: E_j \to V_j$ is a $U(n_1) \times \cdots \times U(n_{p-1/2})$-bundle, where the n_k depend on j. Hence $[\xi^{(j)}] \in \Omega_*\big(B(U(n_1) \times \cdots \times U(n_{p-1/2}))\big)$. Obviously $\Sigma_1^r J([\xi^{(j)}]) = 0 \in \Omega_{n-1}(Z_p)$. There is also the homomorphism $U(n_1) \to$ $\to U(n_1 + 1)$ sending the matrix α into $\begin{pmatrix} 1 & 0 \\ 0 & \alpha \end{pmatrix}$. This will induce a homomorphism

$$I_*: \Omega_m\big(B(U(n_1) \times \cdots \times U(n_{p-1/2}))\big) \to$$
$$\to \Omega_m\big(B(U(n_1 + 1) \times \cdots \times U(n_{p-1/2}))\big).$$

We can now restate (35.2) as

(38.6) *Let (T, M^n) be an orientation preserving diffeomorphism of period p. Let $V_1, \ldots, V_r$ be the components of the fixed point set, and $\xi^{(j)} : E^{(j)} \to V_j$ the normal bundle to V_j. Then $\Sigma J I_*([\xi^{(j)}]) = [T_1, S^1] [M^n]$ in $\Omega_{n+1}(Z_p)$.*

39. Actions of $Z_p \times Z_p$ without stationary points

We have seen in section 30 that if $(Z_2)^k$ acts without stationary points on M^n, then $[M^n]_2 = 0$. By ROCHLIN's theorem if M^n is oriented then $[M^n] \in 2\,\Omega_n$. The corresponding facts for $(Z_p)^k$, p an odd prime, are more complicated; the additional complexity appears to be related in some obscure manner to the divisibility properties of Pontryagin numbers. In this section we settle the matter for $Z_p \times Z_p$. While we are getting ahead of ourselves to consider it now, it provides a non-trivial application of our methods developed up to this point.

We consider, for each compact Lie group G, the collection of all bordism classes in Ω_n which admit a representative M^n on which G acts, preserving orientation, without stationary points. We denote by $SF_n(G)$ this collection of bordism classes in Ω_n, where SF abbreviates "stationary point free". It is clear that $SF_n(G)$ is a subgroup of Ω_n. Moreover $SF(G) = \Sigma SF_n(G)$ is an ideal in Ω, for if G acts on M^n without stationary points, then it acts on $M^n \times V^m$ without stationary points by $g(x, y) = (gx, y)$. Thus $SF(G)$ is an ideal.

The case $G = Z_p$, p a prime, is easy to settle. For if Z_p acts without stationary points on M^n, then Z_p acts freely on M^n, so by (19.4) we have $[M^n] = p[M^n/Z_p]$. Hence $SF(Z_p) \subset p\,\Omega$, and the opposite inequality is clear so that $SF(Z_p) = p\,\Omega$. In generalizing this remark it is useful to consider it in the following fashion. Let Y^0 consist of p distinct points, and note that Z_p acts freely on Y^0 by cyclic permutation. The ideal in Ω generated by $[Y^0]$ is $SF(Z_p)$.

We consider the case $Z_p \times Z_p = G$. There is clearly an action (not effective) without stationary points of $Z_p \times Z_p$ on Y^0. BOREL has pointed out an action without stationary points of $Z_p \times Z_p$ on $P_{p-1}(C)$, [9]. We denote by T and S the generators of $Z_p \times Z_p$. In homogeneous coordinates we set

$$T[z_1, \ldots, z_p] = [z_2, \ldots, z_p, z_1],$$

$$S[z_1, \ldots, z_p] = [z_1, \varrho z_2, \ldots, \varrho^{p-1} z_p]$$

where $\varrho = \exp(2\pi i/p)$. It is seen that $ST = TS$, and the resulting action of $Z_p \times Z_p$ on the complex space has no stationary points. We shall now prove the following.

(39.1) **Theorem.** *The ideal $SF(Z_p \times Z_p)$, for p an odd prime, is the ideal in Ω generated by $[Y^0]$ and $[Y^{2p-2}]$, where $Y^{2p-2} = P_{p-1}(C)$.*

We begin the proof by recalling the procedure just prior to (35.2). We fix an M^n on which $Z_p \times Z_p$ acts, preserving the orientation, without stationary points. We also fix on M^n a Riemannian metric on M^n for which $Z_p \times Z_p$ acts as a group of isometries. Again T, S denote the generators of $Z_p \times Z_p$.

We consider now $F(T)$, the fixed point set of $T : M^n \to M^n$, and let $V_1, \ldots, V_r$ denote the components of $F(T)$. There is the normal bundle $\xi^{(j)} : E^{(j)} \to V_j$ to V_j in M^n. As in section 38, $\xi^{(j)}$ is a $U(n_1) \times \cdots \times U(n_{p-1/2})$-bundle. We first show

(39.2) *In* $\Sigma \, \Omega_* \big(B(U(n_1) \times \cdots \times U(n_{p-1/2})) \big)$ *the element* $\Sigma_1^r [\xi^{(j)}]$ *is divisible by* p.

Proof. Consider a component V of $F(T)$; since $ST = TS$ we have $S(F(T)) \subset F(T)$ and hence either $S(V) = V$ or $V, S(V), \ldots, S^{p-1}(V)$ are all disjoint. We first consider the case in which $V, S(V), \ldots, S^{p-1}(V)$ are disjoint. Suppose $V = V_1, S(V) = V_2, \ldots, S^{p-1}(V) = V_p$. The map $S : V_1 \to V_2$ induces $S : E^{(1)} \to E^{(2)}$ between the respective normal bundles. Since $ST = TS$ it is seen from (38.4) that

$$
\begin{array}{ccc}
E^{(1)} & \overset{S}{\longrightarrow} & E^{(2)} \\
{\scriptstyle \xi^{(1)}} \downarrow & & \downarrow {\scriptstyle \xi^{(2)}} \\
V_1 & \longrightarrow & V_2
\end{array}
$$

has S an equivalence of $U(n_1) \times \cdots \times U(n_{p-1/2})$-bundles. Hence $\xi^{(1)} \cong \xi^{(2)} \cong \cdots \cong \xi^{(p)}$, and $[\xi^{(1)}] + \cdots + [\xi^{(p)}]$ is divisible by p.

Consider next the case $S(V) = V$. Let $V = V_1$. Note that since $Z_p \times Z_p$ has no stationary points, S acts on V_1 without stationary points. As before $S : V_1 \to V_1$ induces a map $S : E^{(1)} \to E^{(1)}$. It follows from (38.4) that S is a bundle map of the $U(n_1) \times \cdots \times U(n_{p-1/2})$-bundle. There is now a commutative diagram

$$
\begin{array}{ccc}
E^{(1)} & \overset{\mu}{\longrightarrow} & E^{(1)}/S \\
{\scriptstyle \xi^{(1)}} \downarrow & & \downarrow {\scriptstyle \bar{\xi}^{(1)}} \\
V_1 & \overset{\nu}{\longrightarrow} & V_1/S
\end{array}
$$

where μ, ν are quotient maps. Moreover $\bar{\xi}^{(1)}$ is also a $U(n_1) \times \cdots \times U(n_{p-1/2})$-bundle, and μ is a bundle map Suppose $f : V_1/S \to B(U(n_1) \times \cdots \times U(n_{p-1/2}))$ induces the bundle $\bar{\xi}^{(1)}$; then $f\nu$ induces $\xi^{(1)}$. We have identified $[V_1, f\nu]$ with $[\xi^{(1)}]$ in $\Omega_* \big(B(U(n_1) \times \cdots \times U(n_{p-1/2})) \big)$. To show $[\xi^{(1)}]$ divisible by p, we have only to apply the following simple extension of (19.4), whose proof we leave to the reader. It uses, besides the proof of (19.4), Theorem (17.5).

(39.3) *Suppose that* V^m *is a closed oriented manifold on which the group* G *of order* r *acts differentiably, freely and preserving the orientation. Let*

V^m/G *be oriented so that the orbit map* $v: V^m \to V^m/G$ *locally preserves orientation. Let* X *be a* CW*-complex for which all torsion in* $H_*(X; Z)$ *consists of elements of order two. If* $f: V^m/G \to X$, *then* $[V^m, fv] = r[V^m/G, f]$ *in* $\Omega_m(X)$.

We can apply (39.3) to our case since $B(U(n_1) \times \cdots \times U(n_{p-1/2}))$ has no torsion. Then $[\xi^{(1)}] = p[\bar{\xi}^{(1)}]$ is divisible by p and (39.2) is complete.

We recall now from 38.6 that $\Sigma J I_*[\xi^{(j)}] = [T_1, S^1][M^n]$ in $\Omega_{n+1}(Z_p)$. Since $\Sigma[\xi^{(j)}]$ is divisible by p, we have the following.

(39.4) *If* $Z_p \times Z_p$ *acts on* M^n, *preserving orientation, without stationary points, then* $[T_1, S^1][M^n]$ *is divisible by* p *in* $\Omega_{n+1}(Z_p)$.

We have now reduced (39.1) to a problem involving the Ω-module structure of $\Omega_*(Z_p)$.

(39.5) *If* $[T_1, S^1][M^n]$ *is divisible by* p *in* $\Omega_{n+1}(Z_p)$, *then* $[M^n]$ *is in the ideal of* Ω *generated by* $[Y^0]$ *and* $[Y^{2p-2}]$.

Proof. Again Y^0 is p points and $Y^{2p-2} = P_{p-1}(C)$. We obtain this result directly from (35.4). Consider a Milnor base $\{[Y^{4k}]\}$ for Ω/Tor with $Y^{2p-2} = P_{p-1}(C)$ as in (36.4). We can write

$$[M^n] = [M_1^n] + [M_2^n] + p[M_3^n]$$

where $[M_1^n]$ is a polynomial in $[Y^{4k}]$ with $4k \neq 2p - 2$ and where $[M_2^n]$ is in the ideal generated by $[P_{p-1}(C)]$. Now from (35.2), $[T_1, S^1][P_{p-1}(C)]$ is divisible by p. Hence $[T_1, S^1][M_1^n]$ must also be divisible by p. From (35.4) this is possible only if $[M_1^n]$ itself is divisible by p; therefore $[M^n] = [M_2^n] + p[M_4^n]$, and (39.5) follows. We have completed the proof of (39.1).

We might note a corollary of (39.1). BOREL has shown that every action of $Z_3 \times Z_3$ on the Cayley plane has a stationary point [6, p. 237]. It follows from (39.1) that every differentiable action of $Z_3 \times Z_3$ on a closed oriented manifold bordant to the Cayley plane has a stationary point. We need the Borel-Hirzebruch results for the Pontrjagin numbers of the Cayley plane [7, p. 535], and we have to compute from MILNOR's results that the Cayley plane is a generator of $\Omega/3\Omega$.

40. Fixed point sets with trivial normal bundles

We consider orientation preserving diffeomorphisms $T: M^n \to M^n$ of odd prime period. It is our contention that there are interesting connections between the geometric properties of such T and the Ω-module structure of $\Omega_*(Z_p)$. In this section we shall study a simple case of this connection.

Here we suppose that *all components of the fixed point set* F *have the same dimension* $m < n$. The normal bundle $\xi: E \to V^m$ over each component is a $U(n_1) \times \cdots \times U(n_{p-1/2})$ bundle as we saw in section 38.

We shall suppose in this section that the numbers $n_1, \ldots, n_{p-1/2}$ are independent of the component of F, and that in $\Omega_m\big(B(U(n_1) \times \cdots \times U(n_{p-1/2}))\big)$ each ξ is bordant to the constant $U(n_1) \times \cdots \times U(n_{p-1/2})$-bundle over V^m. We shall say in this case that F^m has a *trivial normal bundle*.

We have an example of such a T at the beginning of the proof of (36.1). There we considered $T: P_{p-1}(C) \to P_{p-1}(C)$ given by $T[z_1, \ldots, z_p] = [z_1, \varrho z_2, \ldots, \varrho^{p-1} z_p]$, $\varrho = \exp(2\pi i/p)$. This T has p isolated fixed points; in our present notation at each of these points the normal bundle is a $U(1) \times \cdots \times U(1)$-bundle. For any V^m let $T': P_{p-1}(C) \times V^m \to P_{p-1}(C) \times V^m$ be given by $T'(x, y) = (T(x), y)$. Then the fixed point set of T' is p copies of V^m, and the normal bundle is trivial. We shall show that any example is similar to this one.

(40.1) **Theorem.** *Let $T: M^n \to M^n$ be an orientation preserving diffeomorphism of odd prime period p on the closed oriented manifold M^n. Suppose the fixed point set is an m-manifold, F^m, with trivial normal bundle. Then $[F^m] \in p^{a+1}\Omega$ where $(a-1)(2p-2) < n-m \leq a(2p-2)$. If $n-m \neq a(2p-2)$ then $[M^n] \in p\,\Omega$, while if $n-m = a(2p-2)$ then $[M^n] = b\,[P_{p-1}(C)]^a\,[\overline{F}^m]$ in $\Omega/p\,\Omega$ where $[F^m] = p^a\,[\overline{F}^m]$ and $b \neq 0 \bmod p$.*

Proof. By the assumption of trivial normal bundles, there are integers $n_1, \ldots, n_{p-1/2}$ with $n_1 + \cdots + n_{p-1/2} = (n-m)/2$ and with the normal bundle $\xi: E \to F^m$ bordant to a product $U(n_1) \times \cdots \times U(n_{p-1/2})$-bundle over F^m. Then

$$J: \Omega_m\big(B(U(n_1) \times \cdots \times U(n_{p-1/2}))\big) \to \Omega_{n-1}(Z_p)$$

maps $[\xi]$ into $[T, S^{m-n-1}]\,[F^m]$ where S^{n-m-1} is the unit sphere in $C^{n_1} \times \cdots \times C^{n_{p-1/2}}$ and $T: S^{n-m-1} \to S^{n-m-1}$ has $T(v) = \varrho^k v$ for $v \in C^{n_k}$. But $J([\xi]) = 0$, and hence $[T, S^{n-m-1}]\,[F^m] = 0$.

In the Ω-module $\Omega_*(Z_p)$, we must therefore know the annihilator in Ω of $[T, S^{n-m-1}]$. This requires a slight extension of (36.1).

(40.2) *If $[T, X^{2n-1}]$ is an element of $\tilde{\Omega}_{2n-1}(Z_p)$ with $\mu([T, X^{2n-1}]) \neq 0 \in H_{2n-1}(Z_p; Z)$, then the annihilator of $[T, X^{2n-1}]$ is the ideal $p^{a+1}\Omega$ where $a(2p-2) < 2n-1 < (a+1)(2p-2)$.*

Proof. We shall proceed by induction on $2n-1$. It follows by (34.5) and (34.6) for $n = 1$. It also follows from (36.1) that the annihilator contains $p^{a+1}\Omega$. Suppose the lemma is proved for $2n-3$. If $[T, X^{2n-1}] \times [M^m] = 0$, then $(\varDelta_1[T, X^{2n-1}])\,[M^m] = 0$ also. Hence if $a(2p-2) + 3 \leq 2n-1 < (a+1)(2p-2)$ then $[M^m] \in p^{a+1}\Omega$ by induction and the assertion follows for $2n-1$. Suppose next that $2n-1 = a(2p-2) + 1$. Then by induction we have $[M^m] \in p^a\,\Omega$, say $p^a[V^m] = [M^m]$. Then

$$p^a[T, X^{a(2p-2)+1}]\,[V^m] = 0$$

$$b\,[T_1, S^1]\,[P_{p-1}(C)]^a\,[V^m] = 0\,.$$

by (36.2) where $b \neq 0 \bmod p$. By (36.5), $b\,[P_{p-1}(C)]^a\,[V^m] \in p\,\Omega$. Now

$\Omega/p\,\Omega$ is a polynomial algebra and hence is without zero divisors. Therefore $[V^m] \in p\,\Omega$ and $[M^n] \in p^{a+1}\,\Omega$. This completes (40.2).

We return to the proof of (40.1). If $(a-1)\,(2p-2) < n - m \leq \leq a\,(2p-2)$ we have that $[F^m] \in p^a\,\Omega$. With (38.6) we have

$$[T, S^{n-m+1}]\,[F^m] = [T_1, S^1]\,[M^n]$$

for a suitable (T, S^{n-m+1}). Let $[F^m] = p^a\,[\overline{F}^m]$, then

$$p^a\,[T, S^{n-m+1}]\,[\overline{F}^m] = [T_1, S^1]\,[M^n]\,.$$

If $n - m \neq a\,(2p-2)$, then $p^a\,[T, S^{n-m+1}] = 0$ by (36.1). Then $[T_1, S^1]\,[M^n] = 0$ and $[M^n] \in p\,\Omega$ by (36.5). If $n - m = a\,(2p-2)$ then $p^a\,[T, S^{a\,(2p-2)+1}] = b\,[T_1, S^1]\,[P_{p-1}(C)]^a$ by (36.2), where $b \neq 0 \bmod p$. Then $b\,[P_{p-1}(C)]^a\,[\overline{F}^m] = [M^n]$ in $\Omega/p\,\Omega$. Thus (40.1) is proved.

By all odds the simplest case of (40.1) is $p = 3$. Then $(p-1)/2 = 1$ and the normal bundle $\xi : E \to F^m$ becomes a $U(k)$-bundle with $k = (n-m)/2$. The requirement of (40.1) that F^m have trivial normal bundle is considerably simplified over the case $p \neq 3$. In particular, if $m = 0$ the requirements are automatically satisfied, and we get the following:

(40.3) **Corollary.** *Suppose that* $T : M^n \to M^n$ *is an orientation preserving diffeomorphism of period 3 with a finite number of fixed points; then in* $\Omega/3\,\Omega$, $[M^n]$ *is in the polynomial subalgebra generated by* $[P_2(C)]$. *If* $[M^n] \notin 3\,\Omega$ *then there are at least* 3^{a+1} *fixed points where* $4a < n \leq \leq 4\,(a+1)$.

41. Manifolds all of whose Pontrjagin numbers are divisible by p

In this section we consider the ideal in Ω, $I(p) = \Sigma I_n(p)$, where $I_n(p)$ consists of those bordism classes all of whose Pontrjagin numbers are divisible by p, where p is an odd prime. We settle the structure of $I(p)$ completely, using the techniques of MILNOR [26].

According to MILNOR [25, 41] there is in each dimension $4k$ a closed oriented $4k$-manifold Y^{4k} with

$$s_k([Y^{4k}]) = \begin{cases} 1 \text{ if } 2k+1 \text{ not prime power} \\ q \text{ if } 2k+1 = q^r,\ q \text{ a prime} \end{cases}$$

Such a manifold we call a Milnor base element. Our purpose now is to show

(41.1) **Theorem.** *For each odd prime* p, *there exist Milnor base elements* Y^{2p^k-2}, $k = 1, 2, \ldots$ *with all Pontrjagin numbers of* Y^{2p^k-2} *divisible by* p. *The ideal* $I(p)$ *is the ideal generated by* Y^0 *and the* Y^{2p^k-2}, $k = 1, 2, \ldots$, *where* Y^0 *is the 0-manifold consisting of* p *points*.

We suppose first that such Milnor base elements exist. Note that the ideal generated by Y^0 is precisely $p\,\Omega$. Since Ω has no odd torsion, $p\,\Omega$

contains the torsion subgroup T of Ω. Fix Milnor base elements Y^{2p^k-2} as above; fill these out arbitrarily to obtain a Milnor base $[Y^{4i}]$, $i = 1, 2, \ldots$.

We note that the ideal $J(p)$ generated by Y^0 and the Y^{2p^k-2} is contained in $I(p)$. We must prove the converse inclusion. Suppose $[M^n] \in$ $\in I_n(p)$; then

$$[M^n] = \Sigma a_{i_1 \ldots i_k}[Y^{4i_1} \times \cdots \times Y^{4i_k}] + [V^n]$$

where $[V^n] \in J(p)$ and no $a_{i_1 \ldots i_k}[Y^{4i_1} \times \cdots \times Y^{4i_k}]$ is in $J(p)$. Suppose always that $i_1 \geqq \cdots \geqq i_k$ and order the $(i_1, \ldots, i_k)$ with $a_{i_1 \ldots i_k} \neq 0$ lexicographically. Consider the largest $j_1, \ldots, j_k$; then

$$s_{j_1 \ldots j_k}[M^n] = a_{j_1 \ldots j_k} s_{j_1}[Y^{4j_1}] \ldots s_{j_k}[Y^{4j_k}] = 0 \bmod p .$$

If $a_{j_1 \ldots j_k} = 0 \bmod p$ then the term would not appear. Hence $a_{j_1 \ldots j_k} \neq$ $\neq 0 \bmod p$ and so $s_{j_r}[Y^{4j_r}] = 0 \bmod p$ for some r. Then $4j_r = 2p^s - 2$ for some s, and $a_{j_1 \ldots j_k}[Y^{4j_1} \times \cdots \times Y^{4j_k}] \in J(p)$. It follows that $[M^n] \in J(p)$.

It follows that we only need to prove the existence of the Milnor base elements described above. Let $I(p) = \Sigma I_n(p)$. We prove the following via a straightforward use of the methods of MILNOR [26].

(41.2) *For* $n \neq 4k$, $\Omega_n/I_n(p) = 0$. *Moreover*, $\Omega_{4k}/I_{4k}(p) \cong Z_p^{d(k)}$, *where* $d(k)$ *is the number of partitions of* $2k$ *into even integers, none of which is of the form* $p^j - 1$.

Proof. We follow MILNOR in using the Adams spectral sequence for the homotopy groups of the Thom spectrum MSO, [26]. There is a filtration

$$\Omega_n = \{S^0, MSO\}_n = B^{0,n} \supset B^{1,n+1} \supset \cdots$$

and a spectral sequence $\{E_r^{s,t}, d_r\}$ with $E_\infty^{s,t} = B^{s,t}/B^{s+1,t+1}$, and with

$$E_2^{0,t} = \mathrm{Hom}_A^t[H^*(MSO; Z_p), H^*(S^0; Z_p)] ,$$

the Steenrod algebra homomorphisms which lower degree by t. The above cohomology groups are taken to be reduced. MILNOR has shown the spectral sequence is trivial. It also follows from MILNOR's constructions that

$$E_2^{0,n} = 0, \; n \neq 4k, \; E_2^{0,4k} = (Z_p)^{d(k)} .$$

We next show that $I_n(p) = B^{1,n+1}$. Consider a map $f: S^{m+n} \to$ $\to MSO(m)$ which represents a given element of $\Omega_n = \{S^0, MSO\}_n$. There is $f^*: H^*(MSO(m): Z_p) \to H^*(S^{m+n}; Z_p)$, which can be interpreted as an A-homomorphism $H^*(MSO; Z_p) \to H^*(S^0; Z_p)$ lowering degree by n. That is, we can interpret f^* as an element of $E_2^{0,n}$ $= \mathrm{Hom}_A^n[H^*(MSO; Z_p), H^*(S^0; Z_p)]$. We assume the fact that the edge homomorphism of the spectral sequence maps $\{f\} \in \{S^0, MSO\}_n$ into the homomorphism f^*. Now $B^{1,n+1}$ consists of all $\{f\}$ for which f^* is trivial.

A use of the Thom diagram [40] shows $B^{1,n+1}$ to be all $[M^n] \in \Omega_n$ such that, in an embedding of M^n in S^{m+n} (m large), the Pontrjagin numbers of $\mathrm{mod}\, p$ of the normal bundle to M^n are all zero. Use of the Whitney sum theorem for $\mathrm{mod}\, p$ Pontrjagin classes then shows $B^{1,n+1}$ to be all $[M^n]$ whose Pontrjagin numbers (of the tangent structure) are zero $\mathrm{mod}\, p$. Hence

$$\Omega_n / I_n(p) = \Omega_n / B^{1,n+1} \cong E_2^{0,n}$$

and (41.2) is proved.

(41.3) $I_{4k}(p)/p\, \Omega_{4k} \cong (Z_p)^{d'(k)}$, where $d'(k)$ is the number of partitions of $2k$ into even integers at least one of which is of the form $p^j - 1$.

Proof. According to MILNOR [25, 41], $\Omega/p\, \Omega$ is a polynomial algebra over Z_p with one generator from each dimension $4k$. The remark now follows from (41.2) and

$$(\Omega/p\, \Omega)/(I(p)/p\, \Omega) = \Omega/I(p) \; .$$

We now proceed to prove the existence of the Milnor base elements Y^{2p^k-2}. For $k = 1$, $P_{p-1}(C)$ is such a base element. Suppose Y^{2p^l-2} exists for $l \leq k$. There is the ideal $I'(p)$ generated by Y^0, $Y^{2p-2}, \ldots,$ Y^{2p^k-2}. It is seen from (41.3) that

$$I_n'(p) = I_n(p), \; n < 2p^{k+1} - 2 \; ,$$

while $I_n'(p)$ is strictly contained in $I_n(p)$ for $n = 2p^{k+1} - 2$. Let $[M^n]$, $n = 2p^{k+1} - 2$, denote an element of $I(p)$ which is not in $I'(p)$. We show that

$$s_{n/4}[M^n] = ap$$

where $a \neq 0 \, \mathrm{mod}\, p$. Suppose on the other hand that $s_{n/4}[M^n] = bp^2$. Let $[Y^n]$ be a class with $s_{n/4}[Y^n] = p$. Then

$$[M^n] - bp\,[Y^n] = \Sigma a_{i_1 \ldots i_k}\,[Y^{4i_1} \times \cdots \times Y^{4\,i_k}] + \text{torsion}$$

and every non-zero term on the right has $k \geq 2$.

A repetition of the argument used in the first step of the proof of (41.1) shows that in every term there is an r for which $4i_r = 2p^s - 2$. Naturally $s \leq k$. Hence $[M^n] - bp\,[Y^n] \in I'(p)$, and $[M^n] \in I'(p)$ contrary to the hypothesis. Hence

$$s_{n/4}[M^n] = ap, \quad a \neq 0 \, \mathrm{mod}\, p \; .$$

There exist integers c and d with

$$ac + p^k d = 1 \; .$$

Consider now $[c\,M^n + d\,P_{p^{k+1}-1}(C)]$. We have that $c\,[M^n] + d\,[P_{p^{k+1}-1}(c)] \in$ $\in I(p)$. Moreover

$$s_{n/4}(c\,[M^n] + d\,[P_{p^{k+1}-1}(C)]) = acp + dp^{k+1} = p \; .$$

Then $c\,[M^n] + d\,[P_{p^{k+1}-1}(C)]$ serves as $Y^{2p^{k+1}-2}$.

42. Fixed point sets with trivial normal bundles; the general case

Again we consider an orientation preserving diffeomorphism $T: M^n \to M^n$ of odd prime period. Denote by V^m an m-dimensional component of the fixed point set F. The normal bundle $\xi: E \to V^m$ is then a $U(n_1) \times \cdots \times U(n_{p-1/2})$-bundle, as shown in section 38, for appropriate $n_1, \ldots, n_{p-1/2}$. *We shall suppose that the numbers $n_1, \ldots, n_{p-1/2}$ only depend on the dimension m, and not on the component V^m; that there are no components of F in dimension n, and finally that for each V^m, the bundle $[\xi]$ is bordant to the product bundle over V^m in $\Omega_m\big(B(U(n_1) \times \cdots \times U(n_{p-1/2}))\big)$.* We shall say, then, if all these conditions are satisfied by F^m for each m, $0 \leq m < n$, that *the fixed point set of (T, M^n) has trivial normal bundle*. In this section we shall compute the ideal consisting of all bordism classes admitting a representative M^n on which there is an orientation preserving diffeomorphism whose fixed point set F has a trivial normal bundle. It turns out to be just the ideal, $I(p)$, of bordism classes all of whose Pontrjagin numbers are divisible by p. This result has implications about the Ω-module structure of $\Omega_*(Z_p)$. We need a general lemma.

(42.1) **Lemma.** *For a space X, let $\alpha_1, \ldots, \alpha_r$ be homogeneous bordism classes in $\Omega_*(X)$ and suppose that*

$$\Omega_*(X) \xrightarrow{\mu} H_*(X; Z) \xrightarrow{i} H_*(X; Z_p)$$

maps $\alpha_1, \ldots, \alpha_r$ into linearly independent elements of $H_*(X; Z_p)$. If $[M^{n_1}], \ldots, [M^{n_r}] \in \Omega$ are such that $\Sigma \alpha_k [M^{n_k}] = 0$ in $\Omega_*(X)$, then the Pontrjagin numbers of each $[M^{n_k}]$ are all divisible by p.

Proof. Suppose that α_k is represented by a map $f_k : V^{m_k} \to X$. There is the projection $\pi_k : V^{m_k} \times M^{n_k} \to V^{m_k}$. By hypothesis, $\Sigma [V^{m_k} \times M^{n_k}, f_k \pi_k] = 0 \in \Omega_*(X)$. Note that we may as well suppose $m_k + n_k = $ constant. It follows as in Chapter II that if $c \in H^*(X; Z_p)$ and if p_ω denotes a product of Pontrjagin classes, taken mod p, then

$$\Sigma \langle p_\omega(V^{m_k} \times M^{n_k}) \, \pi_k^* f_k^*(c), \, \sigma(V^{m_k} \times M^{n_k}) \rangle = 0$$

in Z_p, where σ denotes the orientation class.

Suppose now that $m_1 \geq \cdots \geq m_r$. We prove the result by induction, assuming that $M^{n_1}, \ldots, M^{n_{k-1}}$ all have Pontryagin numbers $0 \bmod p$. There is a $c \in H^{m_k}(X; Z_p)$ with

$$\langle c, i\mu(\alpha_k) \rangle = 1, \quad \langle c, i\mu(\alpha_l) \rangle = 0, \quad k \neq l.$$

By the additivity theorems for mod p Pontryagin classes

$$p_\omega(V^{m_i} \times M^{n_i}) = 1 \otimes p_\omega(M^{n_i}) + \Sigma a_j \otimes b_j$$

where $\deg a_j > 0$. Finally $\pi_i^* f_i^*(c) = f_i^*(c) \otimes 1$, and the b_j are products of the Pontryagin classes of M^m. Now

$$\langle p_\omega \cdot \pi_l^* f_l^* (c), \sigma\rangle = \langle f_l^* (c) \otimes p_\omega(M^n), \sigma\rangle + \Sigma \langle a_j' \otimes b_j, \sigma\rangle$$

where $\deg a_j' > m_k$. Now

$$\sigma(V^{m_l} \times M^{n_l}) = \sigma(V^{m_l}) \times \sigma(M^{n_l}),$$

so that

$$\langle p_\omega \cdot \pi_l^* f_l^* (c), \sigma\rangle = \langle f_l^* (c), \sigma(V^{m_l})\rangle \langle p_\omega(M^{n_l}), \sigma(M^{n_l})\rangle +$$
$$+ \Sigma \langle a_j', \sigma(V^{m_l})\rangle \langle b_j, \sigma(M^{n_l})\rangle.$$

For $l < k$ the above is zero, by the induction hypothesis about Pontryagin numbers. For $l > k$ we have

$$\langle f_l^* (c), \sigma(V^{m_l})\rangle = \langle c, i\,\mu(\alpha_l)\rangle = 0,$$

while $\langle a_j', \sigma(V^{m_l})\rangle = 0$ since $\deg a_j' > m_l$. We thus see that

$$0 = \Sigma \langle p_\omega \cdot \pi_l^* f_l^* (c), \sigma\rangle$$
$$= \langle p_\omega \cdot \pi_k^* f_k^* (c), \sigma\rangle$$
$$= \langle c, i\,\mu(\alpha_k)\rangle \langle p_\omega(M^{n_k}), \sigma(M^{n_k})\rangle$$
$$= \langle p_\omega(M^{n_k}), \sigma(M^{n_k})\rangle \bmod p.$$

Now (42.1) follows.

(42.2) *If $T : M^n \to M^n$ is an orientation preserving diffeomorphism of period p for which the fixed point set F has a trivial normal bundle, then all the Pontrjagin numbers of M^n are divisible by p.*

Proof. Let $\xi_m : E_m \to F^m$ be the normal bundle, considered as a $U(n_1) \times \cdots \times U(n_{p-1/2})$-bundle. Now by (38.6), $\Sigma J I_*([\xi_m]) = [T_1, S^1] [M^n]$. But by the assumption that the normal bundle is trivial, $J I_*([\xi_m]) = [T, S^{n-m+1}]$ for a suitable periodic map on S^{n-m+1}, and hence $\Sigma_{m<n} [T, S^{n-m+1}] [F^m] - [T_1, S^1] [M^n] = 0$. Now the $i\mu([T, S^{n-m+1}])$ and $i\mu([T_1, S^1])$ are linearly independent in $H_*(Z_p; Z_p)$, so by (42.1) the Pontryagin numbers of $[F^m]$, $0 \leq m < n$ and of $[M^n]$ are divisible by p.

We now set out on the converse problem of (42.2). Namely, given an element of Ω_n all of whose Pontrjagin numbers are divisible by p we must show a representative M^n and an orientation preserving $T : M^n \to M^n$ of period p, such that the normal bundle of the fixed point set is trivial in the sense of this section. In view of section 41, it suffices for each k to find an M^n, $n = 2p^k - 2$, with $s_{n/4}[M^n] = p \bmod p^2$.

The examples will be iterated complex projective space bundles. We first summarize the basic facts concerning these, due to BOREL [5] and BOREL-HIRZEBRUCH [7, p. 513]. Let $\xi : E \to X$ be a $U(n)$-bundle with fiber complex space C^n. There is the unit sphere bundle $\bar\xi : B \to X$, and the action of S^1 (the center of $U(n)$) on B given by scalar multiplication in each fiber. The generated map $\eta : B/S^1 \to X$ is the projective space bundle, with fiber $P_{n-1}(C)$, corresponding to ξ.

Recall that $H^*(B(S^1); Z)$, the cohomology ring of the classifying space of S^1, is a polynomial ring. Denote by $a \in H^2(B(S^1); Z)$ its generator. The characteristic homomorphism $\varrho : H^*(B(S^1); Z) \to H^*(B/S^1; Z)$ maps a in the element $\varrho(a)$, which we also denote by a. Since $\varrho : H^*(B(S^1); Z) \to \to H^*(P_{n-1}(C); Z)$ is an epimorphism the fiber of η is totally non-homologous to zero in B/S^1. Moreover $\eta^* : H^*(X; Z) \to H^*(B/S^1; Z)$ is a monomorphism and every element of $H^*(B/S^1; Z)$ is uniquely represented as

$$\eta^*(x_0) + a\,\eta^*(x_1) + \cdots + a^{n-1} \cdot \eta^*(x_{n-1}) \,.$$

Suppose next that ξ, and therefore also η, is a differentiable bundle. We need here only the case in which the Chern class of ξ is factorable as $(1 + b_1) \ldots (1 + b_n)$. There is the tangent bundle along the fiber in B/S^1; according to BOREL-HIRZEBRUCH [7, p. 514], it has Chern class

(42.3) $(1 - a + \eta^*(b_1)) \ldots (1 - a + \eta^*(b_n))$. But this bundle is actually a $U(n - 1)$-bundle; hence

(42.4) $(a - \eta^*(b_1)) \ldots (a - \eta^*(b_n)) = 0$.

We also need the following remark in order to compute the Chern classes of certain bundles that we encounter.

(42.5) *Suppose that (S^1, X) denotes a free action of S^1 on X, and that the corresponding principal S^1-bundle $X \to X/S^1$ has characteristic Chern class $a \in H^2(X/S^1; Z)$. Consider now $X \times C^k$ with the action of S^1 on the product given by $t(x, z_1, \ldots, z_k) = (t(x), t^{-n_1} \cdot z_1, \ldots, t^{-n_k} \cdot z_k)$ where $n_1, \ldots, n_k$ are fixed integers. The complex vector bundle $\xi : X \times C^k/S^1 \to \to X/S^1$ has Chern class $(1 + n_1 a) \ldots (1 + n_k a)$.*

Proof. It is seen that ξ splits into the Whitney sum of line bundles. It is then sufficient to check the assertion for the case

$$\xi : X \times C/S^1 \to X/S^1 \,,$$

where

$$t(x, z) = (t(x), t^{-n} z) \,.$$

Consider $S^1 \subset C$ and replace $X \times C$ by $X \times S^1$. Let $\nu : X \times S^1 \to \to X \times S^1/S^1$ denote the orbit map. There is the map $\varphi : X \to X \times S^1/S^1$ defined by $\varphi(x) = \nu(x, 1)$. It is seen that

$$\varphi(\varrho(x)) = \varphi(x), \; \varrho = \exp(2\pi i/n)$$

and that φ induces a homeomorphism $X/Z_n \cong X \times S^1/S^1$. The circle group S^1/Z_n acts freely on X/Z_n. Identifying S^1/Z_n with S^1 by $t \to t^n$, we see that

$$\varphi : (S^1/Z_n, X/Z_n) \to (S^1, X \times S^1/S^1)$$

is equivariant. We are thus reduced to computing the Chern class of $(S^1/Z_n, X/Z_n)$.

To do this we look at the Gysin sequence

$$\cdots \to H^r(X) \to H^{r-1}(X/S^1) \xrightarrow{\delta} H^{r+1}(X/S^1) \to \cdots$$

where the Chern class is $\delta(1)$; that is, the image of the unit class under $H^0(X/S^1) \to H^2(X/S^1)$. In this Gysin sequence $H^{r-1}(X/S^1)$ is really $H^{r-1}(X/S^1; H^1(S^1))$, where S^1 is the fiber of $X \to X/S^1$. We have

$$
\begin{array}{ccc}
H^1(X) \to & H^0(X/S^1) \to & H^2(X/S^1) \\
\uparrow & \uparrow{\scriptstyle n} & \Big| \cong \\
H^1(X/Z_n) \to & H^0(X/S^1) \to & H^2(X/S^1)
\end{array}
$$

where n occurs since $H^1(S^1/Z_n) \to H^1(S^1)$ has degree n. Now (42.5) will follow.

We now proceed to the construction of our examples. We set the examples up in two ways, one for geometrical insight and the other for computational purposes; consider first the more geometric. *Suppose we are given a differentiable action (τ, M^n) of the circle group S^1 on the closed oriented manifold M^n.* We define two actions τ_1, τ_2 of S^1 on $I^2 \times M^n$ by

$$t(z, x) = (tz, x), \quad \text{for } \tau_1$$

$$t(z, x) = (tz, t(x)), \quad \text{for } \tau_2$$

where t and z represent complex coordinates for S^1 and I^2 respectively.

Restricting to $S^1 \times M^n$ we obtain induced actions $(\tau_1, S^1 \times M^n)$ and $(\tau_2, S^1 \times M^n)$ which we shall show are equivariantly diffeomorphic. Define $\varphi: S^1 \times M^n \to S^1 \times M^n$ by $\varphi(t, x) = (t, t(x))$. It is easy to check that φ is an equivariant diffeomorphism.

Now from the disjoint union $(\tau_1, I^2 \times M^n) \cup (\tau_2, -I^2 \times M^n)$, we form a closed oriented $(n + 2)$-manifold M^{n+2} and a differentiable action τ of S^1 on M^{n+2} by identifying the boundaries $(\tau_1, I^2 \times M^n)$ and $(\tau_2, I^2 \times M^n)$ via φ.

Thus given (τ, M^n) we receive (τ, M^{n+2}). Consider now the singularities of (τ, M^{n+2}). Note that in $(\tau_1, I^2 \times M^n)$, S^1 acts freely on $I^2 \times M^n -$ $- (0 \times M^n)$, and leaves every point of $0 \times M^n$ stationary. Also in the action $(\tau_2, I^2 \times M^n)$, S^1 acts freely on $I^2 \times M^n - (0 \times M^n)$ while the isotropy subgroup at $(0, x)$ is precisely the isotropy subgroup for (τ, M^n) at x. Thus the singularities of (τ, M^{n+2}) are easily catalogued, as well as the normal bundles. The following remark, easily verified, reveals the relevance of this construction to our problem.

(42.6) *Consider a differentiable action (τ, M^n) of S^1 on the closed oriented manifold M^n, and the action (τ, M^{n+2}) constructed above. There* *are the maps* $T : M^n \to M^n$ *and* $T' : M^{n+2} \to M^{n+2}$ *of period* p *given by* $T(x) = \varrho(x)$, $T'(x) = \varrho(x)$, *where* $\varrho = \exp(2\pi i/p)$. *If the fixed point set of* (T, M^n) *has trivial normal bundle, then so does the fixed point set of* (T', M^{n+2}).

We look now at the alternate description of (τ, M^{n+2}). Consider the action of S^1 on $S^3 \times M^n$ given by $t[(z_1, z_2), y] = [(tz_1, tz_2), t^{-1}(y)]$, and let $S^3 \times M^n/S^1$ be the orbit space. There is the action $(\tau', S^3 \times M^n/S^1)$ of S^1 which is induced from $t[(z_1, z_2), y] = [(tz_1, z_2), y]$ on $S^3 \times M^n$.

(42.7) *There is an equivariant diffeomorphism* θ *of* $(\tau', S^3 \times M^n/S^1)$ *onto* (τ, M^{n+2}).

Proof. Consider first the set $A \subset S^3 \times M^n$ given by $\{[(z_1, z_2), y] : |z_1| \leq |z_2|\}$. Define $\theta_1 : A \to I^2 \times M^n$ by $\theta_1([(z_1, z_2), y]) = (z_1/z_2, (z_2/|z_2|) \cdot y)$. Then $\theta_1([(tz_1, tz_2), t^{-1}(y)]) = \theta_1([(z_1, z_2), y])$ and thus θ_1 induces a map $\bar\theta_1 : A/S^1 \to I^2 \times M^n$. Moreover, $\bar\theta_1 : (\tau; A/S^1) \to (\tau_1, I^2 \times M^n)$ is an equivariant diffeomorphism.

Consider next $B \subset S^3 \times M^n$ given by $\{[(z_1, z_2), y] : |z_1| \geq |z_2|\}$. Define $\theta_2 : B \to I^2 \times M^n$ by $\theta_2[(z_1, z_2), y] = (z_1|z_2|^2/|z_1|^2 z_2, (z_1/|z_1|) \cdot y)$. As before there is an induced map $\bar\theta_2 : B/S^1 \to I^2 \times M^n$, and $\bar\theta_2 : (\tau', B/S^1) \to (\tau_2, I^2 \times M^n)$ is an equivariant diffeomorphism. Finally on $A \cap B$ we have $\varphi \theta_1 = \theta_2$ where $\varphi : (\tau_1, S^1 \times M^n) \cong (\tau_2, S^1 \times M^n)$. There is generated an equivariant diffeomorphism θ of $S^3 \times M^n/S^1$ onto M^{n+2}, and the remark follows.

We shall sometimes denote (τ, M^{n+2}) by $E(\tau, M^n)$. Iterating the function E, we get actions $(\tau, M^{n+2k}) = E^k(\tau, M^n)$. *Thus from* (τ, M^n) *we get a sequence of manifolds* M^{n+2k}, $k = 0, 1, 2, \ldots$ *together with actions of* S^1 *on* M^{n+2k}.

Repeated use of (42.7) yields an explicit formula for M^{n+2k}; namely,

$$M^{n+2k} \cong ((S^3)^k \times M^n)/T^k$$

where the k-dimensional toral group T^k acts by

(*) $(t_1, \ldots, t_k)\,((z_1, w_1), \ldots, (z_k, w_k), y)$

$$= ((t_1 z_1, t_1 w_1), (t_1^{-1} t_2 z_2, t_2 w_2), \ldots, (t_{k-1}^{-1} t_k z_k, t_k w_k), t_k^{-1}(y)).$$

There is the action of T^k on $(S^3)^k$ obtained from (*) by deleting y. Let $L^{2k} = (S^3)^k/T^k$. There is the characteristic homomorphism

$$\varrho : H^*(B(T^k); Z) \to H^*(L^{2k}; Z).$$

Consider $H^*(B(T^k); Z) = Z[a_1, \ldots, a_k]$, and denote the characteristic classes $\varrho(a_i)$ also by a_i.

(42.8) *The cohomology ring* $H^*(L^{2k}; Z)$ *is generated by two dimensional elements* $a_1, \ldots, a_k$ *and these are subject only to the relations* $a_1^2 = 0$, $a_j^2 = a_j a_{j-1}$, $2 \leq j \leq k$. *In particular* $H^{2k}(L^{2k}; Z)$ *is generated by* $(a_k)^k = a_1 a_2 \ldots a_k$.

Proof. For $k = 1$, $L^2 = S^2$ and the remark is clear. Suppose now the result is valid for $k - 1$. We let $T^k = T^{k-1} \times S^1$. There is the projection $(S^3)^k \to (S^3)^{k-1}$ given by $(x_1, \ldots, x_k) \to (x_1, \ldots, x_{k-1})$, and it is seen to be equivariant with respect to the T^{k-1}-actions. There is then induced

$$\xi : (S^3)^k/T^{k-1} \to (S^3)^{k-1}/T^{k-1} = L^{2k-2} ,$$

an S^3-bundle. It is seen that ξ is the sphere bundle of a complex plane bundle $\xi : E \to L^{2k-2}$ where $E = (S^3)^{k-1} \times C^2/T^{k-1}$ is defined just as was L^{2k}. Now $(S^3)^k/T^k = ((S^3)^k/T^{k-1})/(T^k/T^{k-1})$ is just the projective line bundle associated with ξ. We receive then $\eta : L^{2k} \to L^{2k-2}$, with fibre $P_1(C) = S^2$.

We must know the Chern class of ξ. We have introduced (42.5) for this purpose; it shows that $c(\xi) = 1 + a_{k-1}$. We now apply (42.3) and (42.4). We have that $\eta^* : H^*(L^{2k-2}) \to H^*(L^{2k})$ is a monomorphism, that every element of $H^*(L^{2k})$ has a unique representation as $\eta^*(x_0) + a \cdot \eta^*(x_1)$, and from (42.4) that $a(a - a_{k-1}) = 0$. We note that a is the characteristic class a_k. Hence by induction we obtain the remark. We now come to our main assertion.

(42.9) Theorem. *Consider the action (τ, M^{2p-2}) where M^{2p-2} is complex projective space $P_{p-1}(C)$ and where τ is the action of S^1 on M^{2p-2} given by $t[z_1, \ldots, z_p] = [z_1, tz_2, \ldots, t^{p-1}z_p]$. We obtain manifolds $M^{2p+2k-2}$, $k \geq 0$. The manifolds M^n, for $n = 2p^j - 2$ have $s_{n/4}[M^n] = p \bmod p^2$, and hence are Milnor base elements of the bordism algebra $\Omega/p\,\Omega$.*

Proof. Consider $P_{p-1}(C) = S^{2p-1}/S^1$ in the usual fashion. We may regard $M^{2p+2k-2}$ as $(S^3)^k \times S^{2p-1}/T^{k+1}$, where T^{k+1} acts as

$$(t_1, \ldots, t_{k+1})\, ((z_1, w_1), \ldots, (z_k, w_k), (x_1, \ldots, x_p))$$
$$= ((t_1 z_1, t_1 w_1), \ldots, (t_{k-1}^{-1} t_k z_k, t_k w_k), (t_{k+1} x_1, t_k^{-1} t_{k+1} x_2, \ldots, t_k^{-p+1} t_{k+1} x_p))$$

Now then we have a $(2p - 1)$-sphere bundle

$$\xi : (S^3)^k \times S^{2p-1}/T^k \to L^{2k}$$

and $\eta : M^{2p+2k-2} \to L^{2k}$ is the associated complex projective space bundle. It is then seen from our discussion of projective space bundles that $\eta^* : H^*(L^{2k}) \to H^*(M^{2p+2k-2})$ is a monomorphism and that every element of $H^*(M^{2p+2k-2})$ is uniquely represented as

$$\eta^*(x_0) + a_{k+1}\eta^*(x_1) + \cdots + a_{k+1}^{p-1}\eta^*(x_{p-1}) ,$$

where $a_1, \ldots, a_{k+1}$ denote the characteristic classes of the T^{k+1}-action.

It is seen by (42.5) that the Chern class of ξ is

$$(1 + a_k)(1 + 2a_k) \ldots (1 + (p-1)a_k).$$

The tangent bundle μ to $M^{2p+2k-2}$ splits as $\mu = \mu_1 + \mu_2$, where μ_1 is the tangent bundle along the fiber and μ_2 is the normal bundle to the fiber. By (42.3) the Chern class of μ_1 is

$$c(\mu_1) = (1 - a_{k+1})\,(1 - a_{k+1} + \eta^*(a_k)) \ldots (1 - a_{k+1} + (p-1)\,\eta^*(a_k))$$

and by (42.4)

$$a_{k+1}(a_{k+1} - \eta^*(a_k)) \ldots (a_{k+1} - (p-1)\,\eta^*(a_k)) = 0\,.$$

We shall denote a_{k+1} by b and $\eta^*(a_k)$ by a. We see from (42.8) that $a^k b^{p-1}$ is a generator of $H^{2p+2k-2}(M^{2p+2k-2})$, while for $j < p-1$, $a^{p+k-j-1}b^j = 0$ by dimensional considerations. The formulas above become

(i) $\quad b(b-a) \ldots (b - (p-1)a) = 0,$

(ii) $\quad c(\mu_1) = (1-b)\,(1-b+a) \ldots (1 - b + (p-1)a).$

Let $n = 2p^k - 2$, and consider M^n. Recall now the universal Pontryagin class $s_{n/4}(p_1, p_2, \ldots)$. For a vector space bundle τ, let $s_{n/4}(\tau) = s_{n/4}(p_1(\tau), p_2(\tau), \ldots)$. If the base space of τ is an oriented manifold V^n, let $s_{n/4}([\tau]) = \langle s_{n/4}(\tau), \sigma(V^n) \rangle$. Recall also the additivity theorem of THOM [41]

$$s_{n/4}(\tau_1 + \tau_2) = s_{n/4}(\tau_1) + s_{n/4}(\tau_2)$$

modulo 2-torsion.

We now have

$$s_{n/4}[M^n] = s_{n/4}[\mu] = s_{n/4}[\mu_1] + s_{n/4}[\mu_2]\,,$$

and

$$
\begin{aligned}
s_{n/4}[\mu_2] &= \langle s_{n/4}(p_1(\mu_2), \ldots), \sigma(M^n) \rangle \\
&= \langle \eta^*\big(s_{n/4}(p_1(L^{2k}), \ldots)\big), \sigma(M^n) \rangle \\
&= \langle s_{n/4}(p_1(L^{2k}), \ldots), \eta_*(\sigma(M^n)) \rangle \\
&= 0\,.
\end{aligned}
$$

Hence $s_{n/4}\{M^n\} = s_{n/4}[\mu_1]$. It follows from (ii) that the total Pontryagin class of μ_1 is

$$(1 + b^2)\,(1 + (b-a)^2) \ldots \big(1 + (b - (p-1)a)^2\big)$$

and

$$s_{n/4}(\mu_1) = b^{p^k-1} + (b-a)^{p^k-1} + \cdots + (b - (p-1)a)^{p^k-1}\,.$$

We shall now show that

$$s_{n/4}(\mu_1) = p\,a^{p^k-p}b^{p-1} \bmod p^2\, H^n(M^n; Z)\,.$$

Since $a^{p^k-p}b^{p-1}$ is a generator of $H^n(M^n)$, it will then follow that $s_{n/4}[M^n] = p \bmod p^2$.

Since by (i)

$$0 = b(b-a) \ldots (b - (p-1)a) = b^p - a^{p-1}b \bmod p$$

we have

$$b^p = a^{p-1}b + pc \quad \text{for some } c\,.$$

Then,

$$b^{p^2-1} = (b^p)^{p-1}b^{p-1}$$

$$= (a^{p-1}b + pc)^{p-1}b^{p-1}$$

$$= (a^{p^2-2p+1}b^{p-1} + p(p-1)a^{p^2-3p+2}b^{p-2}c)b^{p-1} \bmod p^2$$

$$= a^{p^2-3p+2}b^{2p-3}(a^{p-1}b + p(p-1)c)$$

$$= a^{p^2-3p+2}b^{p-3}(a^{p-1}b + pc)(a^{p-1}b + p(p-1)c)$$

$$= a^{p^2-p}b^{p-1} \bmod p^2 .$$

Similarly we leave to the reader that

$$b^{p^k-1} = a^{p^k-p}b^{p-1} \bmod p^2 .$$

If $r = 1, \ldots, p-1$, then

$$(b - ra)^p = b^p - r^p a^p \bmod p$$

$$= a^{p-1}b - ra^p \bmod p$$

$$= a^{p-1}(b - ra) \bmod p .$$

From the above computation,

$$(b - ra)^{p^k-1} = a^{p^k-p}(b - ra)^{p-1} \bmod p^2 .$$

Then

$$s_{n/4}(\mu_1) = a^{p^k-p}(b^{p-1} + \cdots + (b - (p-1)a)^{p-1}) \bmod p^2 =$$

$$= p\,a^{p^k-p}b^{p-1} \bmod p^2$$

and the theorem (42.9) is proved.

(42.10) **Theorem.** *The ideal in Ω whose homogeneous elements admits representatives M^n on which there is an orientation preserving diffeomorphism $T : M^n \to M^n$ of period p for which the normal bundle to the fixed point set is trivial is precisely the ideal $I(p)$ consisting of those bordism classes all of whose Pontrjagin numbers are divisible by p.*

Proof. Half of our theorem follows from (42.2). Consider $M^{2p-2} = P_{p-1}(C)$ as in Theorem (42.9). There is the action of S^1 on M^{2p-2}; according to the proof of (36.1), the map $T : M^{2p-2} \to M^{2p-2}$ given by $T(x) = \varrho(x)$, $\varrho = \exp(2\pi i/p)$ has trivial normal bundle as defined in this section. We use now (42.9), (42.6) and the results of section 41 to prove the theorem.

Our main reason for making the above constructions is to gain additional knowledge concerning the module structure of $\tilde{\Omega}_*(Z_p)$. We consider generators $\{[T, S^{2n-1}]\}$ for $\tilde{\Omega}_*(Z_p)$, and let $\gamma_k = [T, S^{2n-1}]$. We ask how many, and what are relations of the form

$$\gamma_1 \cdot [M^{n_1}] + \gamma_2 \cdot [M^{n_2}] + \cdots = 0\,?$$

We shall see that the examples of (42.10) give a host of such relations.

Consider the sequence of manifolds $M^0 = p$ points, $M^{2p-2} = P_{p-1}(C)$, $M^{2p}, \ldots$, as constructed in (42.9). As in (42.10) there is the map $T : M^{2p+2k} \to M^{2p+2k}$ of period p. It is seen inductively that the components of $F(T)$, together with their appropriate orientations, are

$$M^{2p+2k-2}, \ -M^{2p+2k-4}, \ldots, (-1)^k M^{2p-2}, (-1)^{k+1} M^0 .$$

The normal bundles ξ are trivial. Using $\Sigma J(\xi) = 0$, we get a relation

$$(42.11) \qquad [T, S^1] [M^{2p+2k-2}] - [T, S^3] [M^{2p+2k-4}] + \cdots$$

$$+ (-1)^k [T, S^{2k-1}] [P_{p-1}(C)] + (-1)^{k+1} p [T, S^{2p+2k-1}] = 0 ,$$

where $T : S^{2j-1} \to S^{2j-1}, j < p + k$ is given by $T(z_1, \ldots, z_j) = (\varrho z_1, \ldots, \varrho z_j)$. However $T : S^{2p+2k-1} \to S^{2p+2k-1}$ is given by $T(x_1, \ldots, x_{p-1}, z_1, \ldots, z_{k+1})$ $= (\varrho x_1, \ldots, \varrho^{p-1} x_{p-1}, \varrho z_1, \ldots, \varrho z_{k+1})$.
We shall return in section 46 to put these relations in final form.

CHAPTER IX

Actions of finite abelian groups of odd prime power order

We now deal with problems of p odd similar to those of Chapter V for $p = 2$. We lead off in section 43 with differentiable, orientation preserving actions of $(Z_p)^k$, p an odd prime, on closed oriented manifolds V^n. The primary aim is to give existence theorems for stationary points of such actions. Our interest in such problems has been aroused by the work of BOREL [6, 9], although we attack the problem from a different point of view. An example of a corollary of our results is that if V^n has one of its Pontryagin numbers not divisible by p then the action has a stationary point. We go on to note that if a toral group acts on V^n without stationary points then $[V^n]$ represents a torsion element of Ω_n.

We give in section 44 those fragments that we know concerning KÜNNETH formulas for $\Omega_*(X \times Y)$. Then in section 45 we consider differentiable, orientation preserving actions of a finite abelian group G of odd prime power order p^k on V^n. Here our real aim is in studying periodic maps of odd prime power period; that is, the case $G = Z_{p^k}$. We succeed in giving existence theorems for stationary points entirely analogous to those of section 43 for the special case $G = (Z_p)^k$. It must be admitted that the proofs here are quite difficult; it is to be hoped that we have overlooked some simpler proof. A surprise in connection with these theorems is that analogous results for maps of period a power of two are false.

These last two chapters having dealt almost entirely with questions concerning the structure of $\Omega_*(Z_p)$ as an Ω-module, it seems appropriate in the last section 46 to summarize what we know about $\Omega_*(Z_p)$. Of course in Chapter VII we settled completely the additive structure of

$\Omega_*(Z_p)$. We now point out that we know considerable concerning $\Omega_*(Z_p)$ as an Ω-module, but presumably not enough to settle all our questions.

43. Actions of $(Z_p)^k$

We illustrate the methods of this chapter by discussing actions of $(Z_p)^k$, p an odd prime.

Given an action (G, V^n) and a subgroup H of G, denote by $F(H, V^n)$ the set of all $x \in V^n$ with $H x = x$. The *singular set* $S(G, V^n)$ is defined to be the union $\cup F(H, V^n)$, where the union is taken over all subgroups H of G with $H \neq \{1\}$. Clearly $S(G, V^n)$ is the minimal invariant subset such that G acts freely on the complement. Often $S(G, V^n)$ is not a finite disjoint union of submanifolds, but the following lemma provides us with a case in which it is.

(43.1) *Suppose that the abelian group G acts on V^n, and that H is a subgroup of G with $G/H \cong Z_p$, p prime. Suppose also that H acts freely on V^n. Consider the family K of subgroups of G with $G = K \times H$. We have $K \cong Z_p$ for each K. Moreover the family $F(K, V^n)$ is pairwise disjoint and $S(G, V^n) = \cup_K F(K, V^n)$.*

Proof. Consider a subgroup K' of G with $K' \neq \{1\}$ and with $F(K', V^n) \neq \emptyset$. Then $K' \cap H = \{1\}$ since H acts freely. Hence the projection $G \to G/H \cong Z_p$ is a monomorphism on K'. Since $K' \neq \{1\}$, then $K' \cong Z_p$. Since K' maps isomorphically onto G/H, it is also seen that $G = K' \times H$.

It follows now that the $F(K, V^n)$ are pairwise disjoint. For if $F(K_1, V^n) \cap F(K_2, V^n) \neq \emptyset$, then $F(K', V^n) \neq \emptyset$ where K' is spanned by K_1 and K_2. Then $K' \cong Z_p$ and hence $K_1 = K_2$. The remark follows.

Suppose given a differentiable, orientation preserving action of $(Z_p)^k$ on a closed oriented manifold V^n, p an odd prime. Suppose also given a Riemannian metric on V^n invariant under the action. We define inductively a sequence
$$V^n, V^{n+1}, \ldots, V^{n+k}$$
of closed oriented manifolds together with actions of $(Z_p)^k$ on V^{n+1}. Here we consider $(Z_p)^k$ as the group generated by elements $T_1, \ldots, T_k$ with the relations $T_j^p = 1$, $T_i T_j = T_j T_i$. Denote by $(Z_p)^l$ the subgroup spanned by $T_{k-l+1}, \ldots, T_k$.

Suppose that V^{n+l} has been defined, $l < k$, together with a differentiable, orientation preserving action of $(Z_p)^k$ on V^{n+l}. Suppose also a Riemannian metric has been fixed on V^{n+l} invariant under the action. *We also take as an induction hypothesis that $(Z_p)^l$ acts freely on V^{n+l}.* There is an action of $(Z_p)^k$ on the unit cell $I^2 = \{z : |z| \leq 1\}$ of the complex numbers given by

$$T_j(z) = \begin{cases} z & \text{for } j \neq k - 1 \\ \varrho z & \text{for } j = k - 1 \end{cases}$$

where $\varrho = \exp(2\pi i/p)$.

Now $(Z_p)^k$ acts on $I^2 \times V^{n+1}$ via the diagonal action; consider in particular the action $((Z_p)^{1+1}, I^2 \times V^{n+1})$. It is seen that

$$S((Z_p)^{1+1}, I^2 \times V^{n+1}) = 0 \times S((Z_p)^{1+1}, V^{n+1}).$$

Since $(Z_p)^1$ acts freely on V^{n+1}, we are now in the setting of (43.1) with $G = (Z_p)^{1+1}$, $H = (Z_p)^1$. Hence

$$S((Z_p)^{1+1}, V^{n+1}) = \cup_K F(K, V^{n+1})$$

where the right hand side is a finite disjoint union, taken over all K with $(Z_p)^{1+1}$ splitting into the direct product of K and $(Z_p)^1$.

In particular, $S((Z_p)^{1+1}, I^2 \times V^{n+1})$ is now a finite disjoint union of closed submanifolds. There is the product Riemannian metric on $I^2 \times V^{n+1}$ invariant under the action of $(Z_p)^k$. We consider a tubular neighborhood N, of small radius, of $S((Z_p)^{1+1}, I^2 \times V^{n+1})$ in $I^2 \times V^{n+1}$. Since S is invariant under $(Z_p)^k$, so are N and $\dot{N}$. Moreover N inherits an orientation from that of $I^2 \times V^{n+1}$, as then does $\dot{N}$. Moreover the action of $(Z_p)^k$ on $\dot{N}$ preserves this orientation. Clearly $(Z_p)^{1+1}$ acts freely on $\dot{N}$, since we have excised the singular set. *We now let* $V^{n+1+1} = \dot{N}$, *with its natural action of* $(Z_p)^k$.

We thus receive by induction the sequence $V^n, \ldots, V^{n+k}$ of closed oriented manifolds, carrying orientation preserving actions of $(Z_p)^k$. We call the sequence a *free resolution* of $((Z_p)^k, V^n)$.

In the above, V^{n+1+1} was the normal sphere bundle to $0 \times S((Z_p)^{1+1}, V^{n+1})$ in $I^2 \times V^n$. We thus get the following.

(43.2) V^{n+1+1} *is the bundle space of a sphere bundle*

$$\xi : V^{n+1+1} \to \cup_K F(K, V^{n+1})$$

where the union is over all subgroups K of $(Z_p)^{1+1}$ with $(Z_p)^{1+1}$ the direct product of K and $(Z_p)^1$. The map ξ is equivariant with respect to $(Z_p)^k$-actions.

Of course in the above we allow the dimension of the spherical fiber to vary from component to component of $F(K, V^n)$.

(43.3) *Consider an action $((Z_p)^k, V^n)$ without stationary points. In the free resolution $V^n, \ldots, V^{n+k}$ we have $V^{n+k} = \emptyset$.*

Proof. Consider the composed map η of

$$V^{n+1+1} \to S((Z_p)^{1+1}, V^{n+1}) \subset V^{n+1}.$$

According to (43.2), each $\eta(x)$ is then fixed under a subgroup K_{i+1} of $(Z_p)^{i+1}$ with $(Z_p)^{i+1}$ the direct product $K_{i+1} \times (Z_p)^i$. Consider now the composition $\Psi : V^{n+k} \to V^n$ of

$$V^{n+k} \to V^{n+k-1} \to \cdots \to V^{n+1} \to V^n \ .$$

For each $\Psi(x)$, there are subgroups $K_k, \ldots, K_1$ of $(Z_p)^k$ with $K_j \Psi(x) = \Psi(x)$ and each K_{i+1} as above. It is seen inductively that $(Z_p)^{i+1}$ is the direct product $K_{i+1} \times K_i \times \cdots \times K_1$. Then $\Psi(x)$ is fixed by the direct product $K_k \times \cdots \times K_1 = (Z_p)^k$. That is, Ψ maps V^{n+k} into $F((Z_p)^k, V^n)$. The remark follows.

We proceed now to the computation of $[(Z_p)^k, V^{n+k}] \in \Omega_{n+k}((Z_p)^k)$; it turns out to depend only on $[V^n]$. However we need first an understanding of $\Omega_*(T^k)$.

Let $T^k = S^1 \times \cdots \times S^1$, and select a point $x_0 \in S^1$. By a *standard torus* $T^i \subset T^k$ we mean a product $X_1 \times \cdots \times X_k \subset T^k$ where each X_j is either x_0 or S^1. Each such torus receives a product orientation from the orientation of S^1. It follows from (18.1) that the bordism classes $[T^i, \mathrm{id}] \in$ $\in \Omega_i(T^k)$ form a basis for the free Ω-module $\Omega_*(T^k)$. Note that for each standard torus $T^i \subset T^k$ there is a dual torus $T^{k-i} = Y_1 \times \cdots \times Y_k$ where $Y_j = S^1$ iff $X_j = x_0$.

Let $f : V^n \to T^k$ be a map of a closed oriented n-manifold into T^k. For each standard torus $T^i \subset T^k$ there is by (10.4) a differentiable approximation g to f which is transverse regular on T^i. It follows from THOM [40] that $g^{-1}(T^i)$ is a closed oriented $(n + i - k)$-submanifold of V^n. It can also be seen that $[g^{-1}(T^i)] \in \Omega_{n+i-k}$ is independent of the choice of the approximation g; we may thus simply write it as $[f^{-1}(T^i)] \in \Omega_{n+i-k}$.

(43.4) *Let $f : V^n \to T^k$ be a map of the closed oriented manifold V^n into T^k. Then $[V^n, f] = 0$ in $\Omega_n(T^k)$ if and only if for each standard torus $T^i \subset T^k$ we have $[f^{-1}(T^i)] = 0$ in Ω_{n+i-k}.*

Proof. Suppose $[f^{-1}(T^i)] = 0$ for each T^i. We can write

$$[V^n, f] = \Sigma [T^i, \mathrm{id}] [M^{n-k}]$$

by (18.1). Suppose now that we have shown that for each standard torus T^m with $m > i$ that $[M^{n-m}] = 0$. We shall show for a given standard torus T^i that $[M^{n-i}] = 0$.

Now $[V^n, f] = \Sigma [T^i, \mathrm{id}] [M^{n-i}]$, where the sum is over all standard tori of dimension $i \leq I$. Form the pair $(\tilde{V}^n, \tilde{f})$ where $\tilde{V}^n$ is the disjoint union $\cup_{i \leq I} T^i \times M^{n-i}$, and where $\tilde{f}$ maps $T^i \times M^{n-i}$ into T^k by projecting onto $T^i \subset T^k$. Then $[V^n, f] = [\tilde{V}^n, \tilde{f}]$ in $\Omega_n(T^k)$.

We seek now an approximation to $\tilde{f}$ which is transverse regular on T^{k-i}, the dual of T^i. Select $y_0 \in S^1$ with y_0 near x_0 but $y_0 \neq x_0$. Define tori $\tilde{T}^i$ analogous to T^i except that the role of x_0 is played by y_0. There are natural maps $r: T^i \to \tilde{T}^i$, all close to the identity. Let $g = f$ on $T^i \times M^{n-i}$ and $g = rf$ on $T^i \times M^{n-i}$ with T^i distinct from T^i. It can now be seen that $g(T^i \times M^{n-i}) \cap T^{k-i} = \emptyset$ if T^i is distinct from T^i and $i \leqq i$. Since T^i and T^{k-i} intersect orthogonally in the point $(x_0, \ldots, x_0)$ it is seen that g is transverse regular on T^{k-i}, and

$$[g^{-1}(T^{k-i})] = [f^{-1}(T^{k-i})] = [M^{n-i}] = 0 .$$

The lemma then follows by induction on $k - l$.

We can now return to free resolutions. On the k-dimensional torus T^k consider the free action of $(Z_p)^k$ given by

$$T_j(z_1, \ldots, z_k) = (z_1, \ldots, \varrho z_j, \ldots, z_k)$$

where $\varrho = \exp(2\pi i/p)$. This gives a free action of $(Z_p)^k$ on T^k, whose class we denote by $\gamma = [(Z_p)^k, T^k] \in \Omega_k((Z_p)^k)$.

(43.5) **Theorem.** *Given a differentiable, orientation preserving action of $(Z_p)^k$ on the closed oriented manifold V^n, consider a free resolution $V^n, V^{n+1}, \ldots, V^{n+k}$ of $((Z_p)^k, V^n)$. The element $[(Z_p)^k, V^{n+k}]$ of $\Omega_{n+k}((Z_p)^k)$ is given by*

$$[(Z_p)^k, V^{n+k}] = [(Z_p)^k, T^k] \, [V^n]$$

in $\Omega_{n+k}((Z_p))$.

Proof. Denote by τ the action of $(Z_p)^k$ on $T^k \times V^n$ given by $g(x, y) = (gx, y)$. For each $0 \leq l \leq k$, let $(Z_p)^k$ act on the torus T^{k-l} by

$$T_j(z_1, \ldots, z_{k-l}) = \begin{cases} (z_1, \ldots, \varrho z_j, \ldots, z_{k-l}) & \text{if } j \leq k - l \\ (z_1, \ldots, z_{k-l}) & \text{if } j > k - l . \end{cases}$$

We interpret the torus T^0 as consisting of a single point. For each $0 \leqq l \leqq k$ define a free action τ_l of $(Z_p)^k$ on $T^{k-l} \times V^{n+l}$ by $g(x, y) = (gx, gy)$. Here (τ_k, V^{n+k}) is just the free resolution $((Z_p)^k, V^{n+k})$.

We shall first prove that $[\tau^k, T \times V^n] = [\tau_0, T^k \times V^n]$ in $\Omega_*((Z_p)^k)$. There is the projection map $T^k \times V^n \to T^k$, equivariant in both the actions τ and τ_0. There are induced maps of orbit spaces

$$\pi_1: (T^k/(Z_p)^k) \times V^n \to T^k/(Z_p)^k \text{ for } \tau ,$$
$$\pi_2: (T^k \times V^n)/(Z_p)^k \to T^k/(Z_p)^k \text{ for } \tau_0 .$$

We regard π_1 and π_2 as inducing the actions τ and τ_0 from the action $((Z_p)^k, T^k)$. It can be seen that in order to prove $[\tau, T^k \times V^n] = [\tau_0, T^k \times V^n]$ it is sufficient to prove that π_1 and π_2 represent the same element of $\Omega_{n+k}(T^k/(Z_p)^k)$. Now $T^k/(Z_p)^k$ is just a torus, so we may use (43.4). Denote $T^k/(Z_p)^k$ by T'^k.

The standard tori in T'^k have the form $X_1 \times \cdots \times X_k/(Z_p)^k$ where each X_j is either S^1 or the subgroup $Z_p \subset S^1$. Since π_1, π_2 are differentiable fiber maps they are transverse regular on each standard torus in T'^k. Now

$$\pi_1^{-1}(T'^{\iota}) = T'^{\iota} \times V^n\,,$$

$$\pi_2^{-1}(T'^{\iota}) = (X_1 \times \cdots \times X_k) \times V^n/(Z_p)^k\,.$$

If $\iota > 0$ clearly $[\pi_1^{-1}(T'^{\iota})] = 0$. We also know from (19.4) that

$$p^k[\pi_1^{-1}(T'^{\iota})] = [X_1 \times \cdots \times X_k \times V^n]\,.$$

Hence if $\iota > 0$ then $p^k[\pi_2^{-1}(T'^{\iota})] = 0$. Since Ω has no odd torsion then $[\pi_2^{-1}(T'^{\iota})] = 0$. If $\iota = 0$, it is seen that

$$[\pi_1^{-1}(T'^0)] = [\pi_2^{-1}(T'^0)] = [V^n]\,.$$

Applying (43.4), we have $[\tau, T^k \times V^n] = [\tau_0, T^k \times V^n]$.

We shall next show that $[\tau_\iota, T^{k-\iota} \times V^{n+\iota}] = [\tau_{\iota+1}, T^{k-\iota-1} \times V^{n+\iota+1}]$ in $\Omega_{n+k}((Z_p)^k)$. Let

$$B^{n+k+1} = S^1_{(1)} \ldots S^1_{(k-\iota-1)} \times I^2 \times V^{n+\iota}\,.$$

There is an action τ'_ι of $(Z_p)^k$ on B^{n+k+1} entirely analogous to τ_ι except that on S^1 is replaced by I^2. In fact,

$$[\tau'_\iota, B^{n+k+1}] = (-1)^{k-\iota-1}[\tau_\iota, T^{k-\iota} \times V^{n+\iota}]\,.$$

Now

$$S((Z_p)^k, B^{n+k+1}) = S^1_{(1)} \times \cdots \times S^1_{(k-\iota-1)} \times 0 \times S((Z_p)^{\iota+1}, V^{n+\iota})\,.$$

A tubular neighborhood M of $S((Z_p)^k, B^{n+k+1})$ is then given by

$$M = S^1_{(1)} \times \cdots \times S^1_{(k-\iota-1)} \times N$$

where N is a tubular neighborhood of $0 \times S((Z_p)^{\iota+1}, V^{n+\iota})$ in $I^2 \times V^{n+\iota}$. Hence

$$\dot{M} = (-1)^{k-\iota-1} S^1_{(1)} \times \cdots \times S^1_{(k-\iota-1)} V^{n+\iota+1}$$

$$= (-1)^{k-\iota-1} T^{k-\iota-1} \times V^{n+\iota+1}\,.$$

Considering the free action of $(Z_p)^k$ on $B^{n+k+1}\backslash \mathrm{Int}\, M$, we have

$$[(Z_p)^k, (B^{n+k+1}\backslash \mathrm{Int}\, M)^\cdot]$$

$$= (-1)^{k-\iota-1}([\tau_\iota, T^{k-\iota} \times V^{n+\iota}] - [\tau_{\iota+1}, T^{k-\iota-1} \times V^{n+\iota+1}]) = 0\,.$$

Hence finally

$$[\tau, T^k \times V^n] = [\tau_k, V^{n+k}]\,,$$

which is the assertion of the theorem.

(43.6) **Corollary.** *Consider a differentiable, orientation preserving action of $(Z_p)^k$ on the closed oriented manifold V^n. Suppose the action has no stationary points. Then $[V^n]$ annihilates the element $\gamma = [(Z_p)^k, T^k]$ of $\Omega_k((Z_p)^k)$; that is, $\gamma \cdot [V^n] = 0$ in $\Omega_{n+k}((Z_p)^k)$.*

Proof. Consider the free resolution $V^n, \ldots, V^{n+k}$. According to (43.3), $V^{n+k} = \emptyset$. Hence from (43.5), $\gamma \cdot [V^n] = 0$.

(43.7) Corollary. *Consider a differentiable, orientation preserving action of $(Z_p)^k$ on the closed oriented manifold V^n. If there are no stationary points, then the Pontryagin numbers of V^n are all divisible by p.*

Proof. By (43.6), $[V^n]$ annihilates the torus γ in $\Omega_*((Z_p)^k)$. However under the composition

$$\Omega_k((Z_p)^k) \to H_k((Z_p)^k; Z) \to H_k((Z_p)^k; Z_p),$$

γ is seen to map into a non-zero element. The result then follows from (42.1).

(43.8) Corollary. *Consider a differentiable action of the toral group T^k on a closed oriented manifold V^n. If there are no stationary points, then the Pontryagin numbers of V^n are all 0 and hence $[V^n]$ represents a torsion element of Ω_n.*

Proof. Consider $(Z_p)^k \subset T^k$. For p sufficiently large, it is seen that $(Z_p)^k$ acts without stationary points. By (43.7), the Pontryagin numbers of V^n are divisible by p for all large primes p. Hence the Pontryagin numbers of V^n are zero.

A closed subgroup H of a compact connected Lie group G is of maximal rank if it contains a maximal toral subgroup. Following Borel, it is of maximal p-rank if it contains a maximal elementary p-group $(Z_p)^k$. We now obtain the following, precisely of the type of Hopf-Samelson.

(43.9) Corollary. *Suppose that G is a compact connected Lie group, and that H is a closed subgroup. If some Stiefel-Whitney number of G/H is not zero, then H is of maximal p-rank. If G/H is orientable and has some Pontryagin number non-zero $\bmod\, p$ for p a prime, then H is of maximal p-rank. If G/H is orientable and has some Pontryagin number not zero, then H is of maximal rank.*

Proof. Suppose for example that $(Z_p)^k$ is a maximal elementary p-group. Let $(Z_p)^k$ act on G/H via $(t, gH) \to tgH$. This action has a stationary point if and only if some conjugate of $(Z_p)^k$ is contained in H. The results now follows from (30.1), (43.7) and (43.8).

In section 39 we introduced the notation $SF(G) = \Sigma_n SF_n(G)$, where $SF_n(G)$ consists of those classes of Ω_n admitting a representative V^n upon which G acts differentially, preserving the orientation and without stationary points. It follows from Chapter VIII that $SF(Z_p) = p\,\Omega$, while $SF(Z_p \times Z_p)$ is the ideal generated by p and $[P_{p-1}(C)]$. Unfortunately we have not settled the structure of $SF((Z_p)^k)$ for $k > 2$. Consider Milnor base elements $Y^0 = p$, $Y^{2p-2}, \ldots, Y^{2p^k-2}, \ldots$, as in section 41 with all Pontryagin numbers divisible by p. We have now shown that $SF((Z_p)^k)$ is contained in the ideal generated by Y^0,

$Y^{2p-2}, \ldots, Y^{2p^l-2}, \ldots$ We conjecture that $SF((Z_p)^k)$ is the ideal generated by $Y^0, \ldots, Y^{2p^{k-1}-2}$. Note also that we have

$$SF(T^k) \subset \text{torsion ideal of } \Omega .$$

We conjecture this to be an equality also.

44. Künneth formulas

In studying actions of $(Z_p)^k$, it would be most helpful to have complete information on $\Omega_*((Z_p)^k)$. Of course a Künneth formula for the structure of $\Omega_*(X \times Y)$ would be very useful. We do not know how to obtain such a formula in general; this section contains such fragments as we do know.

Recall that in section 6 we have defined a homomorphism

$$\chi : \Omega_p(X) \otimes \Omega_q(Y) \to \Omega_{p+q}(X \times Y) ,$$

with

$$\chi([V^m, f] \otimes [W^n, g]) = [V^m \times W^n, f \times g] .$$

It is easy to see that commutativity holds in

$$
\begin{array}{ccc}
\Omega_{p+q}(X \times Y) \otimes \Omega_r(Z) & & \\
\uparrow & & \downarrow \\
\Omega_p(X) \otimes \Omega_q(Y) \otimes \Omega_r(Z) & & \Omega_{p+q+r}(X \otimes Y \otimes Z) \\
\downarrow & & \uparrow \\
& \Omega_p(X) \otimes \Omega_{q+r}(Y \times Z) . &
\end{array}
$$

Letting Y be a single point, commutativity in the above diagram shows that

$$\chi : \Omega_*(X) \otimes \Omega_*(Z) \to \Omega_*(X \times Z)$$

can be factored through $\Omega_*(X) \otimes_\Omega \Omega_*(Z)$. Here we assume a knowledge of the tensor product of graded modules over a graded ring; for example, see CARTAN [12]. We thus arrive at a homomorphism

$$\chi : \Omega_*(X) \otimes_\Omega \Omega_*(Z) \to \Omega_*(X \times Z),$$

the oriented analogue of the homomorphism of section 8.

(44.1) **Theorem.** *Suppose that Y is a finite CW complex with $\Omega_*(Y)$ a free Ω-module. For each CW complex X, the homomorphism*

$$\chi : \Omega_*(X) \otimes_\Omega \Omega_*(Y) \to \Omega_*(X \times Y)$$

is an isomorphism.

Proof. For X a single point, the result is seen to be true since $\Omega_*(p) = \Omega$. We show next an isomorphism

$$\chi : \Omega_*(I^n, S^{n-1}) \otimes_\Omega \Omega_*(Y) \to \Omega_*(I^n \times Y, S^{n-1} \times Y) .$$

Study of the triple $I^n \times Y$, $S^{n-1} \times Y$, $I_-^{n-1} \times Y$ reveals a boundary isomorphism

$$\Omega_*(I^n \times Y, S^{n-1} \times Y) \xrightarrow{\cong} \Omega_*(S^{n-1} \times Y, I^{n-1} \times Y) \cong \Omega_*(I^{n-1} \times Y, S^{n-2} \times Y).$$

The isomorphism $\Omega_*(I^n, S^{n-1}) \cong \Omega_*(I^{n-1}, S^{n-2})$ when tensored with $\Omega_*(Y)$ yields a commutative diagram

$$\Omega_*(I^n, S^{n-1}) \otimes_\Omega \Omega_*(Y) \to \Omega_*(I^n \times Y, S^{n-1} \times Y)$$
$$\downarrow \cong \qquad\qquad\qquad\qquad \downarrow \cong$$
$$\Omega_*(I^{n-1}, S^{n-2}) \otimes_\Omega \Omega_*(Y) \to \Omega_*(I^{n-1} \times Y, S^{n-2} \times Y) \,.$$

The isomorphism

$$\chi : \Omega_*(I^n, S^{n-1}) \otimes_\Omega \Omega_*(Y) \xrightarrow{\cong} \Omega_*(I^n \times Y, S^{n-1} \times Y)$$

follows by induction on n.

Next we induct on the number of cells of X. Suppose the theorem true if X has no more than $k-1$ cells, and consider now a X with k cells. There exists a closed subcomplex X_1 of X containing all but one of the cells of X.

From the exact triangle

$$\Omega_*(X_1) \longrightarrow \Omega_*(X)$$
$$\nwarrow \qquad \swarrow$$
$$\Omega_*(X, X_1)$$

there results, since $\Omega_*(Y)$ is a free Ω-module, the exact triangle

$$\Omega_*(X_1) \otimes_\Omega \Omega_*(Y) \longrightarrow \Omega_*(X) \otimes_\Omega \Omega_*(Y)$$
$$\searrow \qquad \swarrow$$
$$\Omega^*(X \times Y, X_1 \times Y) \,.$$

We now obtain the commutative diagram

$$\Omega_*(X_1) \otimes_\Omega \Omega_*(Y) \longrightarrow \Omega_*(X) \otimes_\Omega \Omega_*(Y)$$

with maps χ_1, χ_2 into

$$\Omega_*(X_1 \times Y) \longrightarrow \Omega_*(X \times Y)$$
$$\Omega_*(X \times Y, X_1 \times Y)$$
$$\chi_3$$
$$\Omega_*(X, X_1) \otimes_\Omega \Omega_*(Y) \,.$$

Diagramm wird zeichnerisch ergänzt — the diagram will be completed by drawing

We have by induction that χ_1 is an isomorphism. Also (X, X_1) is relatively homeomorphic to (I^n, S^{n-1}) and $(X \times Y, X_1 \times Y)$ to $(I^n \times Y, S^{n-1} \times Y)$. Hence χ_3 is an isomorphism. It now follows from the five lemma that χ_2 is an isomorphism. The theorem then follows for Y finite. The extension to an arbitrary CW complex is left to the reader.

A simple example of (44.1) will eventually be of interest, so we dispose of it now. Namely consider a closed oriented M^k, and consider $\Omega_*(S^1 \times M^k)$. Denote by $\alpha_1 = [S^1, \mathrm{id}] \in \Omega_1(S^1)$ the class represented by the identity map of S^1, and by $\alpha_0 = [x_0, i] \in \Omega_0(S^1)$ the class represented by the inclusion of a point $x_0 \in S^1$ into S^1. Then $\Omega_*(S^1)$ is the free Ω-module generated by α_0 and α_1. According to (44.1), we may use χ to identify $\Omega_*(S^1 \times M^k)$ with $\Omega_*(S^1) \otimes_\Omega \Omega_*(M^k)$, and we do so. Hence given a map $f : V^n \to S^1 \times M^k$ where V^n is a closed oriented manifold, there exist $\beta_n \in \Omega_n(M^k)$ and $\beta_{n-1} \in \Omega_{n-1}(M^k)$ with

$$[V^n, f] = \alpha_1 \otimes \beta_{n-1} + \alpha_0 \otimes \beta_n$$

in $\Omega_n(S^1 \times M^k)$. We wish to have at hand explicit formulas for β_n and β_{n-1}. It is easy to do this, following the style of (43.4).

Let $\pi : S^1 \times M^k \to M^k$ be projection. Define $\beta_n \in \Omega_n(M^k)$ to be $[V^n, \pi f]$. To define β_{n-1}, choose an approximation $f' : V^n \to S^1 \times M^k$ which is transverse regular on the submanifold $x_0 \times M^k \subset S^1 \times M^k$. We shall suppose $f' = f$. Define $V^{n-1} = f^{-1}(x_0 \times M^k) \subset V^n$, and define $\beta_{n-1} = [V^{n-1}, \pi f] \in \Omega_{n-1}(M^k)$.

(44.2) *With the notation as above, we have*

$$[V^n, f] = \alpha_1 \otimes \beta_{n-1} + \alpha_0 \otimes \beta_n$$

in $\Omega_n(S^1 \times M^k)$.

Proof. According to (44.1), every $[V^n, f]$ can be written as $\alpha_1 \otimes \beta'_{n-1} + \alpha_0 \otimes \beta'_n$. A proof can now be obtained precisely in the style of (43.4).

Our immediate interest is in $\Omega_*((Z_p)^k)$, where (44.1) does not apply. We go on to show a weaker result in this case.

(44.3) **Theorem.** *Suppose that $B(Z_p)$ is a classifying space for Z_p, p an odd prime. For any CW complex Y the homomorphism*

$$\chi : \Omega_*(B(Z_p)) \otimes_\Omega \Omega_*(Y) \to \Omega_*(B(Z_p) \times Y)$$

is a monomorphism.

Proof. We shall take a particular $B(Z_p)$. Consider the universal space $E(Z_p)$ as the union of the odd dimensional spheres

$$S^1, S^1 \circ S^1, S^1 \circ S^1 \circ S^1, \ldots$$

considered as joins of circles. Consider $Z_p \subset S^1$ as the pth roots of unity, and let Z_p act on S^1 by complex multiplication and diagonally on $S^{2n-1} = S^1 \circ S^1 \circ \cdots \circ S^1$. We make $E(Z_p)$ into a CW complex by prescribing its skeletons to be

$$Z_p,\ S^1,\ S^1 \circ Z_p,\ S^1 \circ S^1,\ S^1 \circ S^1 \circ Z_p,\ \dots .$$

Inspection shows that $E(Z_p)$ has exactly p k-cells for each k, and these are permuted by Z_p.

There is then the classifying space $B(Z_p) = E(Z_p)/Z_p$, which we denote by L. Let L^k denote the k-skeleton of L. Note that L has exactly one cell in each dimension, so that (L^{k+1}, L^k) is relatively homeomorphic to (I^{k+1}, S^k). Note also that L^{2k-1} is a closed oriented manifold, namely a lens space.

We prove by induction on k that

$$\chi : \Omega_*(L^k) \otimes_\Omega \Omega_*(Y) \to \Omega_*(L^k \times Y)$$

is a monomorphism.

Suppose first that this is true for $k = 2n$. The boundary

$$\partial : \Omega_*(L^{2n+1}, L^{2n}) \to \Omega_*(L^{2n})$$

is trivial, since $\Omega_*(L^{2n+1}, L^{2n})$ is a free Ω-module generated by a manifold without boundary. Hence

$$0 \to \Omega_*(L^{2n}) \to \Omega_*(L^{2n+1}) \to \Omega_*(L^{2n+1}, L^{2n}) \to 0$$

is exact. By right exactness of the tensor product,

$$\Omega_*(L^{2n}) \otimes_\Omega \Omega_*(Y) \to \Omega_*(L^{2n+1}) \otimes_\Omega \Omega_*(Y) \to$$
$$\to \Omega_*(L^{2n+1}, L^{2n}) \otimes_\Omega \Omega_*(Y) \to 0$$

is also exact. Also

$$\partial : \Omega_*(L^{2n+1} \times Y, L^{2n} \times Y) \to \Omega_*(L^{2n} \times Y)$$

is seen to be trivial, since

$$\Omega_*(L^{2n+1} \times Y, L^{2n} \times Y) \cong \Omega_*(L^{2n+1}, L^{2n}) \otimes_\Omega \Omega^*(Y)$$

by the discussion of (44.1). We thus receive the commutative diagram

$$\Omega_*(L^{2n}) \otimes_\Omega \Omega^*(Y) \to \Omega_*(L^{2n+1}) \otimes_\Omega \Omega_*(Y) \to \Omega_*(L^{2n+1}, L^{2n}) \otimes_\Omega \Omega_*(Y) \to 0$$

$$\downarrow{\chi_1} \qquad\qquad \downarrow{\chi_2} \qquad\qquad\qquad \downarrow{\chi_3}$$

$$0 \to \Omega_*(L^{2n} \times Y) \to \Omega_*(L^{2n+1} \times Y) \to \Omega_*(L^{2n+1} \times Y, L^{2n} \times Y) \to 0$$

where χ_3 is an isomorphism and χ_1 is a monomorphism. It follows readily from the diagram that χ_2 is a monomorphism.

We next suppose that it has been proved that

$$\chi : \Omega_*(L^{2n-1}) \otimes_\Omega \Omega_*(Y) \to \Omega_*(L^{2n-1} \times Y)$$

is a monomorphism. In order to proceed, we must show that

$$\partial : \Omega_*(L^{2n}, L^{2n-1}) \to \Omega_*(L^{2n-1})$$

is a monomorphism. Consider the diagram

$$\Omega_{2n}(L^{2n}, L^{2n-1}) \xrightarrow{\ \partial\ } \Omega_{2n-1}(L^{2n-1})$$

$$\downarrow{\mu} \qquad\qquad\qquad\qquad \downarrow{\mu}$$

$$H_{2n}(L^{2n}, L^{2n-1}) \xrightarrow{\ \partial\ } H_{2n-1}(L^{2n-1}) .$$

It is seen that $\partial\mu$ maps a generator α of $\Omega_{2n}(L^{2n}, L^{2n-1})$ into $p\,\beta$ where β is a generator of $H_{2n-1}(L^{2n-1})$. Consider then $\alpha' = \partial\alpha$, and note that μ maps α' into $p\,\beta$. Since L^{2n-1} is a manifold, choose a neighborhood U so that $(L^{2n-1}, L^{2n-1}\backslash U)$ is relatively homeomorphic to (I^{2n-1}, S^{2n-2}). Now $\Omega_*(L^{2n-1}, L^{2n-1}\backslash U)$ is a free Ω-module. It is also seen that $\Omega_{2n-1}(L^{2n-1}) \to \Omega_{2n-1}(L^{2n-1}, L^{2n-1}\backslash U)$ maps α' into p times a generator of $\Omega_*(L^{2n-1}, L^{2n-1}\backslash U)$. Since p is odd and Ω has no odd torsion, it follows finally that α' is not annihilated by any element of Ω (except 0). Hence $\partial : \Omega_*(L^{2n}, L^{2n-1}) \to \Omega_*(L^{2n-1})$ is a monomorphism.

We have then the short exact sequence

$$0 \to \Omega_*(L^{2n}, L^{2n-1}) \to \Omega_*(L^{2n-1}) \to \Omega_*(L^{2n}) \to 0$$

and the exact sequence

$$\Omega_*(L^{2n}, L^{2n-1}) \otimes_\Omega \Omega_*(Y) \to \Omega_*(L^{2n-1}) \otimes_\Omega \Omega_*(Y) \to \Omega_*(L^{2n}) \otimes_\Omega \Omega_*(Y) \to 0.$$

Consider now the commutative diagram

$$\Omega_*(L^{2n}, L^{2n-1}) \otimes_\Omega \Omega_*(Y) \to \Omega_*(L^{2n-1}) \otimes_\Omega \Omega_*(Y) \to \Omega_*(L^{2n}) \otimes_\Omega \Omega_*(Y) \to 0$$

$$\downarrow{\chi_1} \qquad\qquad\qquad\qquad \downarrow{\chi_2} \qquad\qquad\qquad\qquad \downarrow{\chi_3}$$

$$\Omega_*(L^{2n}\times Y, L^{2n-1}\times Y) \to \Omega_*(L^{2n-1}\times Y) \to \Omega_*(L^{2n}\times Y)$$

where χ_1 is an isomorphism and χ_2 is a monomorphism. It follows readily from the diagram that χ_3 is a monomorphism.

We have thus shown that

$$\chi : \Omega_*(L^k) \otimes_\Omega \Omega_*(Y) \to \Omega_*(L^k \times Y)$$

is a monomorphism for every k. It can be seen that

$$\Omega_*(L) \otimes_\Omega \Omega_*(Y) \to \Omega_*(L \times Y)$$

is a monomorphism.

Recall that $\Omega_*((Z_p)^k) \cong \Omega_*\big(B((Z_p)^k)\big) = \Omega_*(B(Z_p) \times \cdots \times B(Z_p))$. We thus obtain a natural homomorphism

$$\Omega_*(Z_p) \otimes_\Omega \cdots \otimes_\Omega \Omega_*(Z_p) \to \Omega_*((Z_p)^k) .$$

(44.4) **Corollary.** *For p an odd prime,*

$$\Omega_*(Z_p) \otimes_\Omega \cdots \otimes_\Omega \Omega^*(Z_p) \to \Omega_*((Z_p)^k)$$

is a monomorphism.

Recall that in section 43 there arose the problem of determining the annihilator of the torus $\gamma \in \Omega_k((Z_p)^k)$. Letting $\alpha_1 = [Z_p, S^1] \in \Omega_1(Z_p)$ and identifying $\Omega_*(Z_p) \otimes_\Omega \cdots \otimes_\Omega \Omega_*(Z_p)$ with a submodule of $\Omega_*((Z_p)^k)$, we have $\gamma = \alpha_1 \otimes \cdots \otimes \alpha_1$. At least the problem of computing the annihilator of γ is thus reduced to a problem concerning the Ω-module $\Omega_*(Z_p)$.

We set up now the application of (44.2) and (44.3) that we use in the next section.

Suppose that $Z_p \times H$ acts differentiably on the closed oriented manifold V^n, preserving the orientation. Suppose also that the restriction of this action to the subgroup H is free. We thus receive an element $\beta_n = [H, V^n] \in \Omega_n(H)$. Consider also $\Omega_*(Z_p)$, and the elements $\alpha_0 = [Z_p, Z_p]$ and $\alpha_1 = [Z_p, S^1]$. In α_0, Z_p acts on itself by left multiplication; in α_1, the chosen generator T acts by $T(z) = \varrho z$.

Consider now the action of $Z_p \times H$ on S^1 where Z_p acts as in α_1 and where H acts trivially. There is the diagonal action of $Z_p \times H$ on $S^1 \times V^n$, which we denote by $(\tau, S^1 \times V^n)$. It is seen that τ is a free action, and we wish to compute $[\tau, S^1 \times V^n] \in \Omega_{n+1}(Z_p \times H)$. This we are not able to do, but we do obtain partial information. By (44.3), we may consider $\Omega_*(Z_p) \otimes_\Omega \Omega_*(H)$ as embedded in $\Omega_*(Z_p \times H)$.

(44.5) *With the notation as above, there exists* $\beta_{n+1} \in \Omega_{n+1}(H)$ *with*

$$[\tau, S^1 \times V^n] = \alpha_1 \otimes \beta_n + \alpha_0 \otimes \beta_{n+1}$$

in $\Omega_{n+1}(Z_p \times H)$.

Proof. Use for the classifying space $B(H)$ a closed oriented manifold which is N-classifying for N large. We may suppose that the action (H, V^n) is induced by a differentiable map $\bar{g} : V^n/H \to B(H)$, arising from an equivariant map $g : V^n \to E(H)$. Under the identification $\Omega_*(H) \cong \Omega_*(B(H))$ of section 19, $[H, V^n]$ is identified with $[V^n/H, \bar{g}]$.

Consider now $S^1 \times V^n/Z_p$, which has a free action of H on it. It contains the space $Z_p \times V^n/Z_p$. That is, we consider (H, V^n) as given by an invariant subset of $S^1 \times V^n/Z_p$, namely $Z_p \times V^n/Z_p$. We may then extend $g : V^n \to E(H)$ to an equivariant $h : S^1 \times V^n/Z_p \to E(H)$, which we may as well suppose differentiable.

Let π denote the projection $S^1 \times V^n \to S^1$, and ν the orbit map $S^1 \times V^n \to S^1 \times V^n/Z_p$. There is then

$$f : S^1 \times V^n \to S^1 \times E(H)$$

given by $f(x) = (\pi(x), h\nu(x))$. Using the product action of $Z_p \times H$ on $S^1 \times E(H)$, it may be checked that f is $Z_p \times H$-equivariant. As in (44.3), we may regard S^1 also as the 1-skeleton of $E(Z_p)$. The induced map of orbit spaces

$$\bar{f} : S^1 \times V^n/Z_p \times H \to S'^1 \times E(H) \, ,$$

where $S'^1 = S^1/Z_p$, is such that under the identification $\Omega_*(Z_p \times H) \cong$
$\cong \Omega_*(B(Z_p) \times B(H))$, $[Z_p \times H, S^1 \times V^n]$ is identified with $[S^1 \times V^n/Z_p \times$
$\times H, \bar{j}\bar{f}]$ with j the inclusion $S'^1 \times B(H) \subset B(Z_p) \times B(H)$.

We now consider the element of $\Omega_{n+1}(S'^1 \times B(H))$ which $\bar{f}$ represents.
Letting x_0 be the appropriate base point of S'^1, we see that

$$\bar{f}^{-1}(x_0 \times B(H)) = Z_p \times V^n/Z_p \times H$$

which has been identified with V^n/H. Also $\bar{j}$ on $\bar{f}^{-1}(x_0 \times B(H))$ is seen
to be identified with $\bar{g} : V^n/H \to B(H)$. Applying (44.2), we see that
there exists $\beta_{n+1} \in \Omega_{n+1}(H)$ with

$$[\tau, S^1 \times V^n] = \alpha_1 \otimes \beta_n + \alpha_0 \otimes \beta_{n+1}$$

in $\Omega_{n+1}(Z_p \times H)$.

45. Actions of groups of odd prime power order

We establish in this section existence theorems for fixed points of
periodic maps T of odd prime power period. The proofs very much
resemble the proofs of section 43 on actions of $(Z_p)^k$, but are more difficult
in that the results of section 44 are used. While we are about it, we make
the proofs for actions of any abelian group G of odd prime power order.

Oddly enough, the result of this section do not hold for maps of
period 2^k, as we show by examples.

We begin by considering a version of the setting of (43.1). Namely,
suppose the finite abelian group $Z_p \times H$ acts differentiably on the closed
oriented manifold V^n, preserving the orientation. Suppose also that the
restriction of the action to H is a free action. Let $Z_p \times H$ act on the unit
cell I^2 of the complex numbers C by letting H act trivially and letting
the chosen generator T of Z_p act via $T(z) = \varrho z$. We obtain the diagonal
action $(Z_p \times H, I^2 \times V^n)$. Using (43.1), the singular set is

$$S(Z_p \times H, I^2 \times V^n) = 0 \times S(Z_p \times H, V^n) = \cup_k 0 \times F(K, V^n)$$

where K ranges over the subgroups of $Z_p \times H$ for which $Z_p \times H$ splits
as the direct product of K and H.

It now follows that $S(Z_p \times H, I^2 \times V^n)$ is a finite disjoint union of
closed submanifolds. We may then take a tubular neighborhood N of
$S(Z_p \times H, I^2 \times V^n)$ in $I^2 \times V^n$, of small radius and with orientation
induced by that of $I^2 \times V^n$. We then have a free action of $Z_p \times H$ on
V^{n+1}, preserving the orientation. For want of a better name, we call
$(Z_p \times H, V^{n+1})$ the *free extension* of $(Z_p \times H, V^n)$.

There is the action of $Z_p \times H$ on $S^1 \times V^n$, the restriction of the action
on $I^2 \times V^n$. Since $Z_p \times H$ acts freely on $W^{n+2} = I^2 \times V^n \backslash \text{Int} N$, and
since

$$(Z_p \times H, W^{n+2}) = (Z_p \times H, S^1 \times V^n) \cup (Z_p \times H, -V^{n+1}),$$

we see that

$$[Z_p \times H,\, S^1 \times V^n] = [Z_p \times H,\, V^{n+1}]$$

in $\Omega_{n+1}(Z_p \times H)$. Now let $\beta_n = [H,\, V^n] \in \Omega_n(H)$. We see now from the above equation joined with (44.4) that there exists $\beta'_{n+1} \in \Omega_{n+1}(H)$ with

$$(45.1) \qquad [Z_p \times H,\, V^{n+1}] = \alpha_1 \otimes \beta_n + \alpha_0 \otimes \beta'_{n+1}$$

in $\Omega_*(Z_p \times H)$, where the notation is that of (44.4).

We can now make the main definition of the section.

Definition. Suppose that G is a finite abelian group of odd prime power order p^k, and that

$$0 \subset H_1 \subset \cdots \subset H_k = G$$

is a sequence of subgroups with $H_{l+1}/H_l \cong Z_p$ for $0 \le l < k$. Let $L_j = H_j/H_{j-1}$, $0 < j \le k$, and choose a specific isomorphism of L_j onto Z_p. Suppose now that G acts differentiably on the closed oriented manifold V^n, preserving the orientation. By a *free resolution* of the action $(G,\, V^n)$ we mean first of all a sequence of closed oriented manifolds

$$V^n,\, V^{n+1},\, \ldots,\, V^{n+k}.$$

Secondly, we require that there exist on each V^{n+l} a differentiable action of $(G/H_l) \times L_l \times \cdots \times L_1$, preserving the orientation. In addition, the following are required.

(a) The action of $L_l \times \cdots \times L_1$ on V^{n+l} shall be free. Note that

$$L_{l+1} \times L_l \times \cdots \times L_1 \subset (G/H_l) \times L_l \times \cdots \times L_1$$

and hence $L_{l+1} \times L_l \times \cdots \times L_1$ acts on V^{n+l} with $L_l \times \cdots \times L_1$ acting freely. Note also that $L_{l+1} \times L_l \times \cdots \times L_1$ acts freely on V^{n+l+1}.

(b) It is required that the action $(L_{l+1} \times \cdots \times L_1,\, V^{n+l+1})$ be a free extension of the action $(L_{l+1} \times (L_l \times \cdots \times L_1),\, V^{n+l})$ in the sense described earlier in this section.

Note in (b) that we had previously identified L_{l+1} with Z_p.

It follows from (b) that there is the sphere bundle map

$$V^{n+l+1} \to S(L_{l+1} \times \cdots \times L_1,\, V^{n+l})$$

which goes along with resolutions, and that this map is equivariant with respect to the $L_{l+1} \times \cdots \times L_1$-actions. We obtain now a map as the composition

$$V^{n+l+1} \to S(L_{l+1} \times \cdots \times L_1,\, V^{n+l}) \subset V^{n+l}.$$

(c) It is required that the map $V^{n+l} \to V^n$ obtained as the composition of

$$V^{n+l} \to V^{n+l-1} \to \cdots \to V^n$$

map V^{n+l} into the set $F(H_l,\, V^n)$ of stationary points of H_l, and that the map $V^{n+l} \to F(H_l,\, V^n)$ be equivariant with respect to G/H_l-actions.

This completes the definition!

Consider now a given action (G, V^n). What must be done to construct a resolution? Suppose that $V^n, \ldots, V^{n+\imath}$ have been constructed, together with actions, so that (a)—(c) hold thus far. Following condition (b), there is just one choice for the space $V^{n+\imath+1}$, namely the free extension of $(L_{\imath+1} \times (L_\imath \times \cdots \times L_1), V^{n+\imath})$. This also fixes the action of $L_{\imath+1} \times \cdots \times L_1$ on $V^{n+\imath}$. There is now also fixed the sphere bundle map $\xi \colon V^{n+\imath+1} \to V^{n+\imath}$. We show that $\mu \colon V^{n+\imath+1} \to V^n$ maps $V^{n+\imath+1}$ into $F(H_{\imath+1}, V^n)$ and that $\mu(gx) = \mu(x)$ for $g \in L_{\imath+1} \times L_\imath \times \cdots \times L_1$. Suppose the corresponding fact has already been proved with $\imath + 1$ replaced by $\imath$. Consider $x \in V^{n+\imath+1}$. By the properties of a free extension, $\xi(x) \in V^{n+\imath}$ is fixed under a subgroup K of $L_{\imath+1} \times L_\imath \times \cdots \times L_1$ with $L_{\imath+1} \times \cdots \times L_1$ the direct product of K and $L_\imath \times \cdots \times L_1$. If $g \in L_{\imath+1} \times \cdots \times L_1$, then $g = k\imath$ where $k \in K$ and $\imath \in L_\imath \times \cdots \times L_1$. Hence

$$\mu(g(x)) = \eta\,\xi(k\imath(x)) = \eta\,\xi(\imath(x)) = \eta\big(\imath(\xi(x))\big) = \eta\,\xi(x) = \mu(x)$$

where $\eta \colon V^{n+\imath} \to V^n$. We leave it to the reader to prove that since $\xi(x)$ is fixed under $K \subset (G/H_\imath) \times L_\imath \times \cdots \times L_1$, then $\eta\,\xi(x) = \mu(x)$ is fixed under $H_{\imath+1}/H_\imath$ considered as operating on $F(H_\imath, V^n)$. Hence $\mu(x)$ is fixed under $H_{\imath+1}$.

All that is left to do is to construct an action of $G/H_{\imath+1}$ on $V^{n+\imath+1}$ so that $V^{n+\imath+1} \to F(H_{\imath+1}, V^n)$ is equivariant with respect to the $G/H_{\imath+1}$-actions. This is the delicate part of the construction of a free resolution; we take care of it in the following theorem.

(45.2) **Theorem.** *Let G be a finite abelian group of odd prime power order p^k. Every differentiable action (G, V^n), preserving the orientation on the closed oriented manifold V^n, possesses a free resolution.*

Proof. The above list of requirements for a free resolution is presumably not sufficiently large to allow us to proceed by induction on $\imath$. For purposes of proof, we therefore add the following requirement.

(d) We require that each $V^{n+\imath}$ be a finite disjoint union of submanifolds $V_i^{n+\imath}$, each invariant under the action of $(G/H_\imath) \times L_\imath \times \cdots \times L_1$. It is furthermore required that for each i there exists a toral group $T \subset L_\imath \times \cdots \times L_1$, with T depending on i, and an extension of $(G/H_\imath) \times L_\imath \times \cdots \times L_1$ to an action of $(G/H_\imath) \times T$ such that the map $V_i^{n+\imath} \to V^n$ has the action of T on $V_i^{n+\imath}$ covering the trivial action of T on V^n.

We now assume given a partial resolution $V^n, \ldots, V^{n+\imath}$, satisfying (a)—(d) as far as it goes. As remarked already, $V^{n+\imath+1}$ is determined. It remains to find a suitable action of $(G/H_{\imath+1})$ on $V^{n+\imath+1}$.

In order to consider $V^{n+\imath+1}$ we must consider $I^2 \times V^{n+\imath} = \cup\, I^2 \times V_i^{n+\imath}$. By the inductive assumptions, there is for each i a toral group

$T \supset L_1 \times \cdots \times L_1$ so that $(G/H_1) \times T$ acts on V_i^{n+1} as in (d). We next put an action of $(G/H_1) \times T$ on I^2. Let T act trivially on I^2. Let $L_{1+1} = H_{1+1}/H_1 \cong Z_p$ act on I^2 with the generator of Z_p acting by $z \to \varrho z$. Put differently, we let L_{1+1} act on I^2 by choosing a homomorphism $L_{1+1} \to S^1 = U(1)$. It follows from the character theory of finite abelian groups that the homomorphism $L_{1+1} \to S^1$ can be extended to a homomorphism $G/H_1 \to S^1$, since $L_{1+1} \subset G/H_1$. We thus obtain an operation of $(G/H_1) \times T$ on I^2. Consider finally the diagonal action $((G/H_1) \times T, I^2 \times V_i^{n+1})$.

We now go on to consider the free extension of $(L_{1+1} \times \cdots \times L_1, V_i^{n+1})$. Consider the singular set

$$S(L_{1+1} \times \cdots \times L_1, I^2 \times V_i^{n+1}).$$

According to our previous discussion, it is the union $\cup\, 0 \times F(K, V_i^{n+1})$, taken over all subgroups K of $L_{1+1} \times \cdots \times L_1$ such that $L_{1+1} \times \cdots \times L_1$ splits into a direct product of K and $L_1 \times \cdots \times L_1$. Note that there are just a finite number of choices for K, and choose such a K.

We next express $F(K, V_i^{n+1})$ as a disjoint finite union

$$F(K, V_i^{n+1}) = \cup\, F_j(K, V_i^{n+1}).$$

Here F_j is obtained by choosing a component D of $F(K, V_i^{n+1})$ and letting $F_j = \cup\, gD$, the union taken over all $g \in (G/H_1) \times L_1 \times \cdots \times L_1$.

We shall now define the pieces V_j^{p+1+1} of V^{n+1+1} needed for condition (d). Namely consider a tubular neighborhood N of $0 \times F_j$ in $I^2 \times V_i^{n+1}$, and let $V_j^{n+1+1} = \mathring{N}$. The tubular neighborhoods are assumed to be of small radius, and taken with respect to a Riemannian metric on V_i^{n+1} which is invariant with respect to the action of $(G/H_1) \times T$.

Note that $K \cong Z_p$. In fact it can be seen that there is a homomorphism $\Psi : L_{1+1} \to L_1 \times \cdots \times L_1$ so that K is the set of points $(x, \Psi(x))$ in $L_{1+1} \times (L_1 \times \cdots \times L_1) \subset L_{1+1} \times T$. The component $D' = 0 \times D$ of $0 \times F(K, V_i^{n+1})$ is then a component of the fixed point set of a map of prime period p.

We may thus use the results of section 38. The normal bundle $\xi(D')$ to D' in $I^2 \times V_i^{n+1}$ is then in a natural way a $U(n_1) \times \cdots \times U(n_{p-1/2})$ bundle, where the n_j depend on D'. Consider $g \in (G/H_1) \times T$ and suppose g maps D' into a component D''. Then g induces a bundle map of $\xi(D')$ onto $\xi(D'')$, and in particular $\xi(D')$ and $\xi(D'')$ are $U(n_1) \times \cdots \times U(n_{p-1/2})$-bundles for the same $n_1, \ldots, n_{p-1/2}$. Let $F_j' = gD'$, for all $g \in (G/H_1) \times T$. The normal bundle $N \to F_j' = 0 \times F_j$ is then a $U(n_1) \times \cdots \times U(n_{p-1/2})$-bundle. Moreover $(G/H_1) \times T$ acts on N as a group of bundle maps.

Denote by T' the center of $U(n_1) \times \cdots \times U(n_{p-1/2})$. Note that T' is a torus $(S^1)^{n_1} \times (S^1)^{n_2} \times \cdots \times (S^1)^{n_p-1/2}$. Now T' acts on the sphere bundle $V^{n+1+1} = \mathring{N}$, and commutes with every bundle map $V_j^{n+1+1} \to$

$\rightarrow V_j^{n+\imath+1}$. Hence $(G/H_\imath) \times T \times T'$ acts on $V_j^{n+\imath+1}$. Consider now the subgroup $K \subset (G/H_\imath) \times T$. Recalling that N is made up of normal bundles to components of a fixed point set of $K \cong Z_p$, we see from section 38 that $K \subset (G/H_\imath) \times T$ acts on N, and hence on $V_j^{n+\imath+1}$, as actions of elements of T'. Otherwise put, there is a $\varphi : L_{\imath+1} \rightarrow T'$ so that in $L_{\imath+1} \times T \times T'$ we have that each $(x, \Psi(x), \varphi(x))$ acts trivially on $V_j^{n+\imath+1}$.

The homomorphisms $\Psi : L_{\imath+1} \rightarrow T$ and $\varphi : L_{\imath+1} \rightarrow T'$ can be extended to homomorphisms $\Psi : G/H_\imath \rightarrow T$ and $\varphi : G/H_\imath \rightarrow T'$. We can now define the action of $G/H_{\imath+1}$ on $V_j^{n+\imath+1}$. First of all, let $G/H_\imath$ act by letting $x \in G/H_\imath$ act as does $(x, \Psi(x), \varphi(x)) \in (G/H_\imath) \times T \times T'$. Then $L_{\imath+1} \subset G/H_\imath$ acts trivially, and we thus obtain an action of $(G/H_\imath)/L_{\imath+1} = G/H_{\imath+1}$.

As we have already seen, it is automatic that $V^{n+\imath+1} \rightarrow V^n$ maps $V^{n+\imath+1}$ into $F(H_{\imath+1}, V^n)$. We have finally to check equivariance of $V^{n+\imath+1} \rightarrow V^n$ with respect to the $G/H_{\imath+1}$-actions. Note that the map $V_j^{n+\imath+1} \rightarrow F(K, V_\imath^{n+\imath})$ has the action of T' covering the trivial action of T' on $V_\imath^{n+\imath}$. Hence the map $V_j^{n+\imath+1} \rightarrow V^n$ has the action of $T \times T'$ covering the trivial action on V^n. Let now $g \in G/H_\imath$, which represents an element of $G/H_{\imath+1}$. The action of g on $V_j^{n+\imath+1}$ now covers the action of $g \times \Psi(g)$ on $V_\imath^{n+\imath}$ which covers the action of g on V^n. It is seen that conditions (a)—(d) now hold for $V^n, \dots V^{n+\imath+1}$. The theorem is then proved.

We can now extend (43.6). As in (43.6), $\gamma = [(Z_p)^k, T^k] \in \Omega_k((Z_p)^k)$ denotes the class of the natural action of $(Z_p)^k$ on the torus T^k.

(45.3) **Theorem.** *Let G be a finite abelian group of odd prime power order p^k, and suppose given a differentiable, orientation preserving action of G on the closed oriented manifold V^n. If there are no stationary points, then $[V^n] \in \Omega_n$ annihilates the element $\gamma = [(Z_p)^k, T^k]$ in $\Omega_*((Z_p)^k)$. That is, $\gamma \cdot [V^n] = 0$ in $\Omega_{n+k}((Z_p)^k)$.*

Proof. Consider a free resolution $V^n, \dots, V^{n+k}$ of the action, the existence of which is guaranteed by (45.2). The sequence $0 \subset H_1 \subset H_2 \subset \cdots \subset H_k = G$ is assumed fixed, as are the isomorphisms $L_{\imath+1} = H_{\imath+1}/H_\imath \cong Z_p$. We then obtain specific isomorphisms $L_\imath \times \cdots \times L_1 \cong (Z_p)^\imath$. For each $1 \leq \imath \leq k$, denote by $\beta_{n+\imath} \in \Omega_{n+\imath}((Z_p)^\imath)$ the $[L_\imath \times \cdots \times L_1, V^{n+\imath}]$. It follows from (44.2) that $\beta_{n+\imath} = \alpha_\imath [V^n]$, where $\alpha_\imath = [Z_p, S^1]$. Since the action (G, V^n) has no stationary points, it follows from requirement (c) of a free resolution that $\beta_{n+k} = 0$. It also follows from (45.1) and the definition of a free resolution that for each $1 \leq \imath < k$ there exists $\beta'_{n+\imath+1}$ in $\Omega_{n+\imath+1}((Z_p)^\imath)$ with

$$\beta_{n+\imath+1} = \alpha_1 \otimes \beta_{n+\imath} + \alpha_0 \otimes \beta'_{n+\imath+1} ,$$

where we consider $\Omega_*(Z_p) \otimes_\Omega \Omega_*((Z_p)^\imath)$ as embedded in $\Omega_*((Z_p)^{\imath+1})$.

Putting these facts together, we shall now prove inductively that

$$\beta_{n+k}= 0,\ \alpha_1 \otimes \beta_{n+k-1}= 0,\ \ldots,\ \alpha_1 \otimes \cdots \alpha_1 \otimes \beta_{n+1}= 0\ .$$

Suppose then that $\alpha_1 \otimes \cdots \otimes \alpha_1 \otimes \beta_{n+l+1}= 0$, where there are $k - l - 1$ terms α_1. It follows from (44.4) that

$$\Omega_*(Z_p) \otimes_\Omega \cdots \otimes_\Omega \Omega_*(Z_p) \otimes_\Omega \Omega_*((Z_p)^{l+1})$$

is embedded isomorphically in $\Omega_*((Z_p)^k)$, and it is in that sense that we interpret the above equation.

We now have

$$\alpha_1 \otimes \cdots \otimes \alpha_1 \otimes \beta_{n+1}+ \alpha_1 \otimes \cdots \otimes \alpha_0 \otimes \beta'_{n+l+1}= 0$$

in $\Omega_*((Z_p)^k)$. But since $\Omega_*(Z_p) \cong \tilde{\Omega}_*(Z_p) \oplus \Omega_*$ in the fashion of reduced bordism, we see that

$$\Omega_*(Z_p) \otimes_\Omega \cdots \otimes_\Omega \Omega_*(Z_p) \otimes_\Omega \Omega_*((Z_p)^l) \subset \Omega_*((Z_p)^k)$$

splits into the direct sum of

$$\Omega_*(Z_p) \otimes_\Omega \cdots \otimes_\Omega \tilde{\Omega}_*(Z_p) \otimes_\Omega \Omega_*((Z_p)^l)$$

and

$$\Omega_*(Z_p) \otimes_\Omega \cdots \otimes_\Omega \Omega_* \otimes_\Omega \Omega_*((Z_p)^l)\ .$$

Since $\alpha_1 \otimes \cdots \otimes \alpha_1 \otimes \beta_{n+1}$ belongs to the first of these groups, and $\alpha_1 \otimes \cdots \otimes \alpha_0 \otimes \beta_{n+1}$ to the second, we thus see that $\alpha_1 \otimes \cdots$ $\ldots \alpha_1 \otimes \beta_{n+1} = 0$.

Hence $\alpha_1 \otimes \cdots \otimes \alpha_1 \otimes \beta_{n+1}= 0$, and $(\alpha_1 \otimes \cdots \otimes \alpha_1)\ [V^n] = 0$ in $\Omega_{n+k}((Z_p)^k)$. The theorem then follows.

Just as in (43.7), we have the following corollary.

(45.4) Corollary. *Suppose that the finite abelian group G of odd prime power order p^k acts differentiably on the closed oriented manifold V^n, preserving the orientation and without stationary points. Then the Pontryagin numbers of V^n are all divisible by p.*

We have promised to show such results false in case $p = 2$. In order to do so, we construct certain maps T of period 4. Consider $T : P_2(C) \to P_2(C)$ given by

$$T(z_1, z_2, z_3) = [\bar{z}_1, -\bar{z}_3, \bar{z}_2]\ .$$

Inspection of T shows that it has a single fixed point $[1, 0, 0]$, and that $[1, 0, 0]$ is also an isolated fixed point of T^2. We thus get an orientation preserving action of Z_4 on $P_2(C)$ with just one stationary point. Moreover Z_4 acts freely on a deleted neighborhood of that point. Taking the diagonal action of Z_4 on $P_2(C) \times P_2(C)$, we also get such an action on $P_2(C) \times P_2(C)$. Finally, we get such an action of Z_4 on $P_4(C)$ by taking T with

$$T(z_1, z_2, z_3, z_4, z_5) = (\bar{z}_1, -\bar{z}_3, \bar{z}_2, -\bar{z}_5, -\bar{z}_4)\ .$$

We get then orientation preserving actions of Z_4 on $P_4(C)$ and on $P_2(C) \times P_2(C)$. Each action has just one stationary point, and Z_4 acts freely on deleted neighborhoods of the stationary point.

From representation theory, taking orientation into account there are precisely two ways in which Z_4 can act orthogonally on an 8-ball with the origin the only singularity. By excising neighborhoods of the stationary points of $P_4(C)$ and $P_2(C) \times P_2(C)$, and fitting the results together along the boundary, we thus get a manifold V^8 upon which Z_4 acts without stationary points. We can choose an orientation so that Z_4 preserves orientation and so that $[V^8] = [P_4(C)] \pm [P_2(C) \times P_2(C)]$. If necessary we can now add two copies of $P_2(C) \times P_2(C)$ to make the sign positive.

(45.5) **Example.** *There exists a differentiable, orientation preserving action of Z_4 on a closed oriented manifold V^8, without stationary points and with*

$$[V^8] = [P_4(C)] + [P_2(C) \times P_2(C)] .$$

Note that V^8 has Pontryagin numbers not divisible by two. For example,

$$s_2[V^8] = s_2[P_4(C)] = 5 .$$

Recall the symbol $SF(G) = \Sigma SF_n(G)$, where $SF_n(G) \subset \Omega_n$ consists of those bordism classes admitting a representative V^n upon which G acts differentiably, preserving the orientation and without stationary points. It appears from the above that $SF(Z_4)$ is rather large. A likely possibility for it is to all bordism classes of even Euler characteristic; that is, all $[V^n]$ with $w_n[V^n] = 0$. On the other hand, for p odd $\Omega_*(Z_{p^k})$ is not nearly so large. All we know is that it is between $p\,\Omega$ and the annihilator of $\gamma \in \Omega_k((Z_p)^k)$. For all we know, it can be that $SF(Z_{p^k}) = p\,\Omega$. A test case here could be $SF(Z_9)$. It can be seen using our techniques that $SF(Z_9)$ contains $3\,\Omega$ and is contained in the ideal generated by 3 and $[P_2(C)]$.

Question. Is there a closed oriented manifold V^4, bordant to $P_2(C)$, upon which there acts a periodic differentiable map of period 9, preserving the orientation and without fixed points.

The examples used in (45.5) also raise a question, in connection with which we make the following conjecture.

Conjecture. There cannot exist a periodic differentiable map of odd prime power period acting on a closed oriented manifold V^n, $n > 0$, preserving the orientation and possessing exactly one fixed point.

46. The module structure of $\Omega_*(Z_p)$

Here we summarize what we know concerning the module structure of $\Omega_*(Z_p)$. Our first theorem continues section 42.

(46.1) *Consider the generating set $\alpha_{2k-1} : k = 1, 2, \ldots$ for $\Omega_*(Z_p)$, p an odd prime, where $\alpha_{2k-1} = [T, S^{2k-1}]$ is given by $T(z_1, \ldots, z_k) = (\varrho z_1, \ldots, \varrho z_k)$. There exist closed oriented manifolds M^{4k}, $k = 1, 2, \ldots$, such that for each k,*

$$p\,\alpha_{2k+1} + [M^4]\,\alpha_{2k-3} + [M^8]\,\alpha_{2k-7} + \cdots = 0$$

in $\Omega_*(Z_p)$.

Proof. We define inductively a sequence $M^2, M^4, \ldots$ of closed oriented manifolds, together with differentiable maps $T : M^{2k} \to M^{2k}$ of period p.

To define M^2, recall that $p\,[T, S^1] = 0$. Hence there exists a closed oriented 2-manifold M^2 and a $T : M^2 \to M^2$ of period p, having exactly p fixed points, each having an oriented neighborhood I^2 in which T is given by $T(z) = \varrho z$.

Suppose now that (T, M^{2k}) has been defined. Consider $(\tau_1, I^2 \times M^{2k})$ and $(\tau_2, I^2 \times M^{2k})$, where τ_1 and τ_2 are actions of Z_p given by $\tau_1(x, y) = (\varrho x, y)$ and $\tau_2(x, y) = (\varrho x, ty)$. Accordingly to (35.2), $[\tau_1, I^2 \times M^{2k}]$ and $[\tau_2, I^2 \times M^{2k}]$ are equal in $\Omega_{2k+1}(Z_p)$. There exists then a differentiable fixed point free action τ of Z_p on a compact oriented B^{2k+2} with

$$(\tau, \dot{B}^{2k+2}) = (\tau_1, I^2 \times M^{2k}) - (\tau_2, I^2 \times M^{2k}) .$$

Define (T, M^{2k+2}) by suitable identification of boundaries in

$$(\tau_1, I^2 \times M^{2k}) - (\tau, B^{2k+2}) - (\tau_2, I^2 \times M^{2k}) .$$

Hence we obtain $M^2, M^4, \ldots$. It is seen inductively that the fixed point set of (T, M^{2k}), together with appropriate orientation, consists of

$$M^{2k-2}, -M^{2k-4}, M^{2k-6}, \ldots, (-1)^{k-1} M^0$$

where $M^0 = p$ points. The normal bundles are trivial in a suitably strong sense, so that

$$\alpha_1 [M^{2k-2}] - \alpha_3 [M^{2k-4}] + \alpha_5 [M^{2k-6}] - \cdots + (-1)^{k-1} p\,\alpha_{2k-1} = 0 .$$

Since Ω_{4i+2} contains only 2-torsion and each α_{2i-1} is of odd order, it is seen that alternate terms of this expression are zero. We thus obtain

$$p\,\alpha_{2k-1} + [M^4]\,\alpha_{2k-5} + [M^8]\,\alpha_{2k-9} + \cdots = 0 .$$

The remark is then proved.

Suppose now for each k we select $M^4, M^8, \ldots$, possibly depending on k, with

$$\beta_{2k-1} = p\,\alpha_{2k-1} + [M^4]\,\alpha_{2k-5} + [M^8]\,\alpha_{2k-9} + \cdots = 0$$

in $\Omega_*(Z_p)$.

(46.2) *Consider the free Ω-module C with generator $\alpha_1, \alpha_2, \alpha_3, \ldots$ and $\partial : C \to C$ given by $\partial \alpha_{2k-1} = 0$ and $\partial \alpha_{2k} = \beta_{2k-1}$. Then $\Omega_*(Z_p) \cong H_*(C)$ as Ω-modules.*

Proof. Define the submodule $C^{(k)} \subset C$ to be the submodule generated by $\alpha_1, \alpha_2, \ldots, \alpha_k$. Then

$$0 \subset C^{(1)} \subset C^{(2)} \subset \cdots \subset C$$

is a filtration of the chain complex C, and there is a spectral sequence $\{'E_{p,q}^r\}$. Also $'E^\infty$ is associated with a filtration of $H_*(C)$. Note that

$$'E_{p,q}^1 = H_{p+q}(C^{(p)}/C^{(p-1)}) \, .$$

A straight-forward analysis shows

$$'E_{2k,q}^2 = 0, \quad 'E_{2k-1,q}^2 = Z_p \otimes \Omega_q \, .$$

Just as with the bordism spectral sequence for $B(Z_p)$, the spectral sequence is trivial. Moreover $H_n(C)$ and $\Omega_n(Z_p)$ are seen to have the same order.

Define a homomorphism $C \to \Omega_n(Z_p)$ given by $\alpha_{2k-1} \to [T, S^{2k-1}]$, $\alpha_{2k} \to 0$. There results a homomorphism $H_*(C) \to \Omega_*(Z_p)$, seen to be an epimorphism. Since $H_n(C)$ and $\Omega_n(Z_p)$ are finite groups of the same order, this must be an isomorphism. The remark follows.

(46.3) Theorem. *Consider the generating set $\alpha_{2k-1} : k = 1, 2, \ldots$ for $\Omega_*(Z_p)$, and closed oriented manifolds $M^{4k}, k = 1, 2, \ldots$, such that for each k*

$$\beta_{2k-1} = p\,\alpha_{2k-1} + [M^4]\,\alpha_{2k-5} + [M^8]\,\alpha_{2k-9} + \cdots = 0 \, .$$

The ideal of Ω generated by p and all the $[M^{2k}]$ coincides with the ideal of all elements of Ω whose Pontryagin numbers are all divisible by p. Moreover $\Omega_*(Z_p)$ is isomorphic as an Ω-module to the quotient of the free Ω-module generated by $\alpha_1, \alpha_3, \ldots$ by the submodule generated by $\beta_1, \beta_3, \ldots$.

Proof. We already have from (46.1) and (46.2) all the conclusions except those dealing with Pontryagin numbers. That the Pontryagin numbers of M^{4k} are all divisible by p follows from (42.1). We shall now prove that M^{2p^k-2} is a Milnor base element for $\Omega/p\,\Omega$; this will conclude the proof.

For each fixed k we shall show that there exist manifolds $V^0, V^4, \ldots$ with

$$(*) \qquad \gamma_{2p^k-1} = [V^0]\,\alpha_{2p^k-1} + \cdots + [V^{2p^k-2}]\,\alpha_1 \in p\,\Omega_*(Z_p)$$

and with V^{2p^k-2} a Milnor base element of $\Omega/p\,\Omega$. Suppose this granted for the moment. According to (46.2)

$$\gamma_{2p^k-1} = p\,\gamma + b\,\beta_{2p^k-1} + [W^4]\,\beta_{2p^k-5} + \cdots,$$

for suitable $[W^{4k}]$ and γ, in the free module generated by $\alpha_1, \alpha_3, \ldots$. Then

$$[V^{2p^k-2}] = p\,[V'^{2p^k-2}] + b\,[M^{2p^k-2}] + [W^4]\,[M]^{2p^k-6} + \cdots .$$

Since $[V^{2p^k-2}]$ is a Milnor base element of $\Omega/p\,\Omega$, then $b \neq 0 \bmod p$ and

$[M^{2p^k-2}]$ is a Milnor base element of $\Omega/p\,\Omega$. Hence we have only to find relations of the type (*). But these follow directly from (42.11).

For an element $\gamma \in \Omega_*(X)$, the annihilator $A(\gamma)$ is the ideal of Ω consisting of all $[M^n]$ with $\gamma \cdot [M^n] = 0$. There has arisen in (43.4) the problem of computing the annihilator $A(\gamma_k)$ of the toral action $\gamma_k \in$ $\in [(Z_p)^k, T^k] \in \Omega_k((Z_p)^k)$. In the notation of section 45, $\gamma_k = \alpha_1 \ldots \alpha_1$ and it follows from section 44 that we may as well compute the annihilator in the submodule $\Omega_*(Z_p) \otimes_\Omega \cdots \otimes_\Omega \Omega_*(Z_p)$. Unfortunately we cannot compute this annihilator, but we have the following.

(46.4) *The annihilator $A(\gamma_{k+1})$ contains the elements p, $[M^{2p-2}]$, ...,* $[M^{2p^k-2}]$ *where the M^{4k} are as in* (46.3).

Proof. The proof is by induction on k. For $k = 0$, the assertion is the known fact that $p\,\alpha_1 = 0$. Suppose it has been proved for $k - 1$, so that the ideal generated by p, $[M^{2p-2}]$, ..., $[M^{2p^{k-1}-2}]$ annihilates γ_k. Now from (46.3),

$$[M^{2p^k-2}]\,\alpha_1 = -[M^{2p^k-6}]\,\alpha_5 - \cdots - p\,\alpha_{2p^k-1}\,.$$

Moreover it follows from section 41 that the coefficients of the right hand side are in the ideal generated by p, $[M^{2p-2}]$, ..., $[M^{2p^{k-1}-2}]$. Then

$$[M^{2p^k-2}]\,\alpha_1 \ldots \alpha_1 = -\alpha_5([M^{2p^k-6}]\,\alpha_1 \ldots \alpha_1) - \cdots$$
$$-\alpha_{2p^k-1}(p\,\alpha_1 \ldots \alpha_1) = 0\,.$$

The remark follows.

We have already conjectured that $A(\gamma_{k+1})$ is precisely the ideal generated by p, $[M^{2p-2}]$, ..., $[M^{2p^k-2}]$, but the proof of the opposite inclusion appears difficult.

List of references

[1] ATIYAH, M. F.: Bordism and cobordism. Proc. Cambridge Philos. Soc. **57**, 200—208 (1961).

[2] AVERBRUCH, V. G.: Algebraic structure of the group of intrinsic homology. Doklady Akad. Nauk SSSR **125**, 11—14 (1959).

[3] BLAKERS, A. L., and W. S. MASSEY: On the homotopy groups of a triad. II. Ann. Math. **55**, 192—201 (1952).

[4] BOREL, A.: La cohomologie mod 2 de certains espaces homogènes. Comment. Math. Helv. **27**, 165—197 (1953).

[5] — Sur la cohomologie des espaces fibrés principaux et des espaces homo-gènes des groupes de Lie compact. Ann. Math. **57**, 115—207 (1953).

[6] — Sous-groupes commutatifs et torsion des groupes de Lie compacts connexes. Tohoku Math. J. **13**, 216—240 (1961).

[7] —, and F. HIRZEBRUCH: On characteristic classes of homogeneous spaces; I. Am. J. Math. **80**, 458—538 (1958); II. Am. J. Math. **81**, 315—382 (1959).

[8] —, and J. P. SERRE: Groupes de Lie et puissances réduites de Steenrod. Am. J. Math. **75**, 409—448 (1953).

[9] — et al.: Seminar on transformation groups. Ann. Study 46, Princeton (1960).

[10] BOURGIN, D. G.: On some separation and mapping theorems. Comment. Math. Helv. 29, 199—214 (1955).

[11] CARTAN, H., and others: Cartan Seminar Notes, 1950—51, Paris.

[12] — Cartan Seminar Notes, 1954—55, Paris.

[13] CONNER, P. E., and E. E. FLOYD: Differentiable periodic maps. Bull. Am. Math. Soc. 68, 76—86 (1962).

[14] — Fixed point free involutions and equivariant maps. Bull. Am. Math. Soc. 66, 416—441 (1960).

[15] DOLD, ALBRECHT: Démonstration élémentaire de deux résultats du cobordism. Ehresmann Seminar notes 1958—59, Paris.

[16] — Erzeugende du Thomschen Algebra $\mathfrak{N}$. Math. Z. 65, 25—35 (1956).

[17] — Vollständigkeit der Wuschen Relationen zwischen den Stiefel-Whitney-schen Zahlen differenzierbarer Mannigfaltigkeiten. Math. Z. 65, 200—206 (1956).

[18] EILENBERG, S.: On the problems of topology. Ann. Math. 50, 247—260 (1949).

[19] —, and N. E. STEENROD: Foundations of Algebraic Topology. Princeton, 1952.

[20] HIRZEBRUCH, F.: Neue topologische Methoden in der algebraischen Geometrie. Berlin-Göttingen-Heidelberg: Springer-Verlag 1956.

[21] MILNOR, J. W.: Differentiable structures on homotopy spheres, mimeographed notes. Princeton, 1959.

[22] — Differential Topology, mimeographed notes, Princeton, 1958.

[23] — Groups which act on S^n without fixed points. Am. J. Math. 79, 623—630 (1957).

[24] — Lectures on Characteristic Classes, mimeographed notes. Princeton, 1958 (to appear in book form).

[25] — On the cobordism ring Ω^*. Notices Am. Math. Soc. 5, 457 (1958).

[26] — On the cobordism ring Ω^* and a complex analogue, Part I. Am. J. Math. 82, 505—521 (1960).

[27] — On the relation between differentiable manifolds and combinatorial manifolds, mimeographed notes. Princeton, 1956.

[28] — Some consequences of a theorem of Bott. Ann. Math. 68, 444—449 (1958).

[29] MONTGOMERY, D., and L. ZIPPIN: Topological Transformation Groups. New York, 1955.

[30] MORSE, A. P.: The behavior of a function on its critical set. Ann. Math. 40, 62—70 (1939).

[31] MOSTOW, G. D.: Equivariant embeddings in Euclidean space. Ann. Math. 65, 432—446 (1957).

[32] PALAIS, R. S.: Imbedding of compact differentiable transformation groups in orthogonal representations. J. Math. and Mech. 6, 673—678 (1957).

[33] PONTRYAGIN, L. S.: Characteristic cycles on differentiable manifolds. Math. Sbornik 21, 233—284 (1947).

[34] ROHLIN, V. A.: Intrinsic homologies. Doklady Akad. Nauk SSSR 89, 789—792 (1953).

[35] SERRE, J. P.: Groupes d'homotopie et classes de groupes abelians. Ann. Math. 58, 258—294 (1953).

[36] SPANIER, E. H.: Infinite symmetric products, function spaces, and duality. Ann. Math. 69, 142—198 (1959).

[37] —, and J. H. C. WHITEHEAD: Duality in homotopy theory. Mathematika 2, 56—80 (1955).

[38] STEENROD, N. W.: Topology of Fibre Bundles. Princeton, 1951.

[39] THOM, R.: Espaces fibrés en sphères et carrés de Steenrod. Ann. Sci. Ecole Norm. Sup. **69**, 109—182 (1952).

[40] — Quelques propriétés globales des variétés différentiables. Comment. Math. Helv. **28**, 17—86 (1954).

[41] — Travaux de Milnor sur le cobordisme. Bourbaki Seminar notes 1958—59, Paris.

[42] WALL, C. T. C.: Determination of the cobordism ring. Ann. Math. **72**, 292—311 (1960).

[43] WHITEHEAD, G. W.: Generalized homology theories. Trans. Am. Math. Soc. **102**, 227—283 (1962).

[44] WHITEHEAD, J. H. C.: Combinatorial homotopy I. Bull. Am. Math. Soc. **55**, 213—245 (1949).

[45] WHITNEY, H.: Differentiable manifolds. Ann. Math. **37**, 645—680 (1936).

[46] WU, WEN-TSÜN: Les i-carrés dans une variété grassmanniene. C. R. Acad. Sci. (Paris) **230**, 918—920 (1950).

[47] YANG, C. T.: Continuous functions from spheres to Euclidean spaces. Ann. Math. **62**, 284—292 (1955)

[48] — On theorems of Borsuk-Ulam, Kakutani-Yamobe-Yujobo and Dyson I. Ann. Math. **60**, 262—282 (1954).